W9-BDC-504

Nonparametric Comparative Statics and Stability

Nonparametric Comparative Statics and Stability

DOUGLAS HALE
GEORGE LADY
JOHN MAYBEE
JAMES QUIRK

PRINCETON UNIVERSITY PRESS
PRINCETON, NEW JERSEY

Published by Princeton University Press, 41 William Street,
Princeton, New Jersey 08540
In the United Kingdom: Princeton University Press,
Chichester, West Sussex

Library of Congress Cataloging-in-Publication Data

Nonparametric comparative statics and stability / by Douglas Hale...
[et al.].
p. cm.
Includes bibliographical references and index.
ISBN 0-691-00690-3 (cloth : alk. paper)
1. Economics, Mathematical. 2. Macroeconomics—Mathematical
models. 3. Equilibrium (Economics). I. Hale, Douglas, 1946–
HB135.N664 1999
330′.01′51--dc21 98-55311
CIP

This book has been composed in Times Roman with Helvetica Display

The paper used in this publication meets the minimum requirements of
ANSI/NISO Z39.48-1992 (R1997) (*Permanence of Paper*)

http://pup.princeton.edu

Printed in the United States of America

1 3 5 7 9 10 8 6 4 2

To Yvonne M. M. Bishop

Two True Scholars: David F. Lady and Dorothy E. Lady

Carol

Shirley

Contents

Preface

The fundamental issue addressed in this book concerns the degree to which subject matter such as economics can be considered a science, i.e., a source of refutable hypotheses. Such a concern is not limited to the realm of scholarly debate, but arrives with ever-increasing importance in an age when many governmental and commercial policies are derived from computer-based renditions of conceptual models from economics and elsewhere.

Our particular experience with applied analysis includes the energy market forecasting systems developed by the Energy Information Administration (EIA) of the U.S. Department of Energy over the decades since the 1973–1974 oil embargo. The issues at stake in the use of these systems include the degree of expected dependency upon foreign sources of energy supply; the effect on energy production and costs of adopting, or removing, the regulation of energy markets; and the consequences of undertaking alternative polices designed to ameliorate the environmental problems associated with the production and consumption of many energy products. The outcomes of policy issues such as these have substantial material significance for all of us; further, the design of many specific policies depends upon exactly how the data about the relationships at issue are used. Given this, the EIA has maintained an aggressive interest in reviewing and determining the operational quality of its policy analysis modeling.

The particular circumstances that led to this book can be traced in part to the EIA model quality control program. In light of the importance of modeling to the energy policy debate, a symposium was held at the University of Colorado in 1980, with attendees invited from a broad range of disciplines (proceedings in Greenberg and Maybee [1981]). The papers and discussions at the symposium covered the technical issues that needed to be resolved, and promising approaches for resolving them, in order to understand the basis for the expected accuracy of the results from large computer-based forecasting systems. Many of the results reported here stem from investigations prompted by this symposium. Some of the examples presented in Chapter 4 were initially sponsored by the EIA as part of its continuing interest in model quality control.

But the subject matter at issue is not limited to energy, or economics, but instead, potentially can by applied to any inquiry. Even in disciplines traditionally viewed as "fully quantitative," there can be problems with the accuracy of measurement and questions about what types of conclusions can be drawn nevertheless. An extensive bibliography on topics in (as called here) nonparametric comparative statics and stability may be found at the end of the book. In reviewing this, the reader can note instances of application in many scientific areas, including artificial intelligence, biology, chemistry, ecology, energy (as already noted), and large physical systems. Further, there has been a continuously active interest in the mathematical underpinnings of the analysis. We believe that some of the frames of reference for the analysis here that seem particular to economics, i.e., optimization and equilibrium, have rather straightforward analogues in other disciplines; e.g., the balance conditions for equilibrium correspond to double-entry concepts in accounting and conservation laws in physics. As a result, although we come to many issues within the framework of our own background in economics, it is hoped that the approaches taken and the results achieved can serve as a resource in the study of similar problems in other disciplines. We believe that the issue of "nonparametrics" must necessarily be present to one degree or another in any applied science.

The technical content of the book made this a very difficult book to edit. The quality of the editing job done by Lyn Grossman cannot be overstated. Production editor Molan Chun Goldstein directed the various tasks involved in actually producing the book. The authors wish to particularly acknowledge this support. Of course the ultimate responsibility for any remaining errors remains with the authors.

This book is specially dedicated to the memory of John Maybee. John was a friend and colleague of each of us for many decades and contributed to the preparation of this manuscript. John passed away before the book was completed. John's characterization of the structure of determinants in terms of cycles and chains is critical for the arguments in the book. He combined a rare ability to make esoteric mathematical arguments accessible to nonspecialists with good-natured humor. He is sorely missed.

Nonparametric Comparative Statics and Stability

1

Nonparametric Analysis

1.1 INTRODUCTION

In the early 1950s, the federal government decided to stockpile strategic materials to protect against the possibility of hostilities with the USSR. Planners constructed a massive 400×400 input-output model of the U.S. economy, entailing the estimation of a very large number of input-output coefficients, mainly based on data from the 1947 Census of Manufactures, to assess the effects of alternative inventory holdings. After several years of data gathering and analysis, they tested the resulting model, using the final bill of goods for the economy for 1951. Among other findings from that test was that the calculated domestic steel requirement for producing the 1951 final bill of goods was 40% more than the capacity of the U.S. steel industry in 1951, a physical impossibility. What had gone wrong was that relative prices of different grades of steel (alloy, stainless, carbon) had changed between 1947 and 1951, leading to a change in the output mix of the steel industry. Price changes in the steel industry had also led to changes in steel-using technology, and substitution of other metals for steel, by the customers of the steel industry. In the space of four years, the adaptiveness of the American economy had clearly revealed the volatility of the input-output coefficients underlying the Defense Department's economic planning models.[1] Problems with the volatility of input-output and other such coefficients continue to be commonplace in economic research.[2]

Since the quantitative particulars of the interrelationships that constitute economic phenomena are so often volatile and transitory, it is natural to inquire about aspects of the interrelationships that might be more stable and robust. This book explores what can be said about the collective outcome of interdependent quantitative phenomena when the precise nature and magnitudes of their separate influences are not known. Although the particular case of economic phenomena motivated the original work in this area, the problem of inference with limited quantitative information is endemic to scientific inquiry. Scientific explanations of observable phenomena are based on structural relations:

$$F(y; \alpha, e) = (f_1, \ldots, f_n),$$

where the f_i are functions linking the phenomena to be explained (the endogenous variables), $y = (y_1 \ldots y_n)$, to conditioning numbers (parameters and exogenous variables, collectively, the data), $\alpha = (\alpha_1, \ldots, \alpha_m)$, determined outside the theory, and unobserved random disturbances, e. The random disturbance, e, is suppressed in most of this book.[3] Scientific predictions are derived under the assumption that observed values of the phenomena, y, are equilibrium values, $y^* = y^*(\alpha, e)$, defined by

$$F(y^*; \alpha, e) = (f_1, \ldots f_n) = 0_n \tag{1.1}$$

Much of the daily work of scientists involves making observations and conducting experiments to estimate the form of the equation systems such as (1.1), the values of conditioning numbers, and the distribution of the unobserved disturbances. The result of a successful research program is a complete and internally consistent explanation of the phenomena. Comparison of its predictions to data can then test the validity of the theory.

1.2 QUANTITATIVE ANALYSIS

Quantitative analysis in economics has traditionally focused on comparative statics: the problem of computing changes in the equilibrium values of endogenous variables induced by changes in the data. Analysis of the local direction of change in economic magnitudes in response to changes in technology, resource endowments, people's preferences, and public policy naturally results in locally linear systems of equations under appropriate differentiability assumptions (see Samuelson 1947). In this book, most of the systems we analyze are local comparative statics models. In recent years, the comparative statics problem has been reformulated by Milgrom and Shannon (1994), Milgrom (1994), and Milgrom and Roberts (1990, 1994). This approach is directed at establishing conditions necessary and sufficient for global qualitative comparative statics results. Because of its generality, the approach is less useful as a device for identifying the specific comparative statics results that economic models can generate. The precise links between this approach and that taken in this book have not been completely established. However, in at least one case, that involving the maximization hypothesis, the correspondence between the approach taken here and the Milgrom-Shannon monotonicity theorem is easily established. See the discussion in Chapter 5.

Comparative Statics

The analysis is initiated by noting that in the neighborhood of an equilibrium, y^*, the changes induced in y^* by changes in α can be written in differential form as

$$\Sigma_j(\partial f_i/\partial y_j)dy_j^* + \Sigma_k(\partial f_i/\partial \alpha_k)d\alpha_k = 0, \tag{1.2}$$

where $i, j = 1, \dots, n$ and $k = 1, \dots, m$. If only one exogenous variable changes, the most common case in economic modeling, the differential system becomes

$$\Sigma_j(\partial f_i/\partial y_j)dy_j^*/d\alpha = -\partial f_i/\partial \alpha \qquad i = 1, \dots, n. \tag{1.3}$$

Determining the change in the equilibrium values of phenomena with respect to a change in an exogenous variable, $dy_j^*/d\alpha$, is the subject matter of comparative statics ("statics" because time does not explicitly enter (1.1)–(1.3). Define the square $n \times n$ matrix, $A = [a_{ij}] = [\partial f_i/\partial y_j]$, the $n \times 1$ vector, $x = (dy_i^*/d\alpha)$, and the $n \times 1$ vector, $b = [b_j] = [-\partial f_i/\partial \alpha]$. The local comparative statics problem then can be written as

$$Ax = b, \tag{1.4}$$

where x is to be determined. The matrix A is called the *Jacobian matrix*, corresponding to a solution to the system (1.3).

A theory is locally scientific in the sense of Popper ([1934] 1959) if for a given, potentially observable b-vector, a particular x-vector could never arise as a solution to (1.4). The theory would be "refuted" if the particular x-vector were in fact observed. From the standpoint of refutable hypotheses, the content of a theory is represented by the characteristics of its Jacobian matrix.

Dynamics

The equilibrium y^* defined by (1.1) is often interpreted as the stationary state associated with a dynamic adjustment process operating on the phenomena, y, over time, t. Formally, the adjustment process is

$$\dot{y} \equiv dy/dt = g(y; \alpha, e) = (g_1, \dots g_n), \tag{1.5}$$

where dy/dt is the time derivative of y. When the rate of change of each y_i is increasing with its "distance" from equilibrium as measured by f_i, the adjustment mechanism can be written as

$$\dot{y} \equiv dy/dt = g(f) = (g_i(f_i)). \tag{1.6}$$

A linear approximation of (1.6) in the neighborhood of y^* is obtained by a first-order Taylor series, $dy_i/dt = (dg_i/df_i)\Sigma_j(\delta f_i/\delta y_j)(y_j - y_j^*)$, evaluated at the equilibrium y^*. Writing this expression in matrix form yields

$$\dot{y} \equiv dy/dt = DA(y - y^*), \tag{1.7}$$

where D is a diagonal matrix with $d_{ii} = dg_i/df_i > 0$ and A is as defined in (1.4). Global stability analysis is concerned with determining conditions that ensure that (1.5) or (1.6) devolves to zero in the limit, for arbitrary initial conditions. Linear approximation stability analysis seeks conditions on D and A ensuring that (1.7) devolves to zero in the limit, in a neighborhood of y^*.[4]

1.3 NONPARAMETRIC ANALYSIS

The quantitative approach to scientific explanation breaks down when the theory's underlying equations, conditioning numbers, and unobserved disturbances are only vaguely known. In the natural sciences, where the relationships are presumed immutable in time and space, nascent theories may reveal the import of novel data sets that may not be available for decades. In the social sciences, both the underlying relationships and the magnitudes of the conditioning numbers may change with place and time. People change, institutions change, and the technology changes: careful observation and estimation may still yield only provisional approximations of a transient reality.[5] The demands of quantitative analysis are often simply not feasible. The validity of plausible inferences from theories, mental models, and computer programs can all be rejected on the basis of quantitative information, but the problem of inference with limited quantitative information remains. Social scientists in particular have little hope of ever achieving precise knowledge of people and their organizations.

Quantitative results in economic applications consequently have limited predictive power. Little is known about the actual form of the underlying relationships, and controlled field experiments to resolve magnitudes are seldom feasible. Even in the linearized structure of (1.4) or (1.7), precise quantitative information is typically absent. Formally, the magnitudes of the entries of A and b in (1.4) and D in (1.7) are not completely known. Given this, the immediate issue becomes, What *can* be safely assumed to be known? For the purposes of this book, any state of knowledge about the nature of (1.4) of (1.7) that is less than fully quantified will be termed *nonparametric*.

A basic difficulty in performing scientific work outside of a fully quantitative environment is that the set of nonparametric information available to researchers can come in a seemingless endless variety of forms. Depending on his or her progress, a researcher may know only which variables appear in individual relationships (i.e., which entries of A in (1.4) are zero and which are not), or the direction of influence of parameters on variables (i.e., the signs $[+, -, 0]$ of the entries of A in (1.4)), or the relative magnitudes of some of the entries of A (i.e., a ranking of the entries of A in (1.4)). There is no single nonparametric environment. Researchers have pursued two related approaches to this curse of riches.

One approach focuses on the types of information that are likely to be available to researchers concerning the entries of the Jacobian matrix. The classic example, developed in Chapter 2, is to assume researchers know only whether entries are greater than, less than, or equal to zero. This emphasis on sign information arose historically in economics because economists are most secure in their beliefs about which variables appear in relationships and the nature of their direct influence, i.e., whether $\partial f_i / \partial y_j$ and $\partial f_i / \partial \alpha$ are zero, positive, or negative. An analysis based upon sign pattern information alone is termed a *qualitative analysis*. Thus, a qualitative analysis deals with the matrices in equations (1.4) and (1.7) under the assumption that sign pattern information is available concerning the Jacobian matrix A and the vector b. Chapter 2 presents the results available for qualitative analyses. Sometimes a researcher may assume to know additional information about the matrix's entries, such as their relative sizes or bounds upon their magnitudes. Chapter 3 organizes sign pattern information analysis with these additional categories of information into a hierarchy analogous to that of measurement scales. Results are derived that show how the different categories of information about the entries of the Jacobian matrix can lead to definitive conclusions about the entries of the inverse Jacobian matrix. The other approach, developed in Chapters 5–8 is to hypothesize underlying principles, such as maximization or stability, governing the Jacobian matrix of the systems described by (1.4) and (1.7). These principles, combined with qualitative information, can sometimes yield definite results.

In all of these the fundamental mathematical questions are the same. First, under what conditions, given the information assumed to be available about the entries of the Jacobian matrix and the vector b (in (1.4)), can we solve (partially or completely) for the sign pattern of the vector x? And second, when can the same information enable us to determine the stability of the differential equation system (1.7)?

1.4 AN EXAMPLE

It might be helpful to consider an example of a qualitative analysis. In a simple full-employment economy, current output, X, is fixed at a level X_f. Total real output is divided among investment, consumption, and government expenditure, all expressed in real (inflation-adjusted) dollars. Equivalently, output can be viewed as the sum of real savings, consumption, and taxes. Investment, I, is assumed to decrease with the interest rate, i. Consumption, C, is assumed to increase with disposable income, X_d, which is defined as output less taxes, i.e., $X_d = X - T$. For the purposes of this example, taxes and government expenditures are assumed to be exogenous variables set by government policy, i.e., they can be chosen independently of other economic variables.

In the money market, it is assumed that the money supply, M is set by the central bank. The real money supply is M/P, where P is the price level. Demand for real money balances consists of transactions demand, kX, where k is a constant and X, as above, is real output; and speculative or liquidity demand $L(i)$, a decreasing function of the interest rate. In equilibrium, the demand for real money balances equals the supply of real balances. The equilibrium equations governing this simply economy, based upon market clearing in the goods and money markets, are

$$G + I(i) + C(X_d) = X$$

and

$$M/P = kX + L(i),$$

where $X = X_f$ and $X_d = X - T$.

Differentiating the two equilibrium equations totally with respect to all endogenous and exogenous variables yields

$$\begin{bmatrix} dI/di & 0 \\ -dL/di & -\dfrac{M}{P^2} \end{bmatrix} \begin{bmatrix} di \\ dP \end{bmatrix} = \begin{bmatrix} -dG + (dC/dX_d)dT \\ -dM/P \end{bmatrix}.$$

The differentials dG, dT, and dM are policy changes selected by the government or the central bank and result in changes in the interest rate and the price level, di and dP, respectively. The signs $(+, 0, -)$ of the derivatives are established by the assumptions made earlier. The qualitative system corresponding to (1.4) is

$$\begin{bmatrix} - & 0 \\ + & - \end{bmatrix} \begin{bmatrix} di \\ dP \end{bmatrix} = \begin{bmatrix} -dG + (dC/dX_d)dT \\ -dM/P \end{bmatrix}.$$

Inverting the coefficient matrix and solving yields

$$\begin{bmatrix} di \\ dP \end{bmatrix} = \begin{bmatrix} - & 0 \\ - & - \end{bmatrix} \begin{bmatrix} -dG + (dC/dX_d)dT \\ -dM/P \end{bmatrix}.$$

Consider what happens if government expenditures are increased ($dG > 0$), while taxes and the money supply are held constant ($dT = dM = 0$):

$$\begin{bmatrix} di \\ dP \end{bmatrix} = \begin{bmatrix} - & 0 \\ - & - \end{bmatrix} \begin{bmatrix} - \\ 0 \end{bmatrix} = \begin{bmatrix} + \\ + \end{bmatrix}.$$

Thus, this system is fully sign solvable—the signs of both di and dP are determined by the sign pattern information given. An increase in real government expenditures in this economic model gives rise to an increase in the interest rate and an increase in the price level, assuming taxes and the money supply are held fixed. Further, it is easy to verify that changing taxes or changing the money supply, holding the other exogenous variables fixed, also leads to a fully sign solvable system. Thus, within this model, increasing government expenditure increases the interest rate and the price level (both di/dG and dP/dG are positive). Increasing taxes reduces both the interest rate and the price level (both di/dT and dP/dT are negative). Increasing the money supply increases prices ($dP/dM > 0$), but does not affect the interest rate ($di/dM = 0$). To say that this simple economy is sign solvable means that, assuming that equilibrium is reestablished, it is possible to deduce the direction of change in all the economic variables as government policy changes, *independent of the magnitudes of the influences expressed by the Jacobian matrix so long as the directions of the influences (i.e., signs of the entries) are those assumed.*

The comparative statics approach assumes that a new equilibrium is established, given the changes in government policy. But in order to guarantee that equilibrium will be reestablished following an exogenous change, the dynamic adjustment mechanism must be stable. The dynamic adjustment mechanism for this economy is based on two assumptions. If investors attempt to invest more than savers save, the interest rate increases. Rising interest rates are assumed to choke off lower return investment opportunities. Similarly, if the supply of money exceeds what people wish to hold, the price level goes up; in effect, too much money is chasing too few goods. In terms of (1.5) the equations are

$$di/dt = b_1(I(i) - S);$$

and

$$dP/dt = b_2(M/P - kX - L(i)),$$

where the b's are positive constants measuring the "speeds of adjustment" of P and I; i.e., the b's determine how rapidly i and P adjust to disturbances to equilibrium. In the linear approximation form of (1.6) the equations become

$$di/dt = b_1\{(dI/di)(i - i^*)\}$$

and

$$dP/dt = b_2\{-(dL/di)(i - i^*) - (M/P^2)(P - P^*)\},$$

where * indicates an equilibrium value. In qualitative terms, the linear differential equation system corresponding to (1.7) becomes

$$\begin{bmatrix} di/dt \\ dP/dt \end{bmatrix} = \begin{bmatrix} + & 0 \\ 0 & + \end{bmatrix} \begin{bmatrix} - & 0 \\ + & - \end{bmatrix} \begin{bmatrix} (i - i^*) \\ (P - P^*) \end{bmatrix}.$$

It can be shown that this is a sign stable system: any matrix with the above sign pattern has both characteristic roots with negative real parts. Thus if equilibrium is disturbed by a policy change in G, T, or M, i and P will converge asymptotically to new equilibrium values, with qualitative changes as described above.

This example captures the analytic goals, and the main topics, of the material presented in this book. A model of phenomena is proposed, but something less that a full specification of its quantitative attributes is assumed to be known. Given this, what can be said about the comparative statics, and stability, of its solution? The example was contrived to present a case for which sign pattern information about the Jacobian matrix was sufficient to determine the sign pattern of its inverse and its stability. For larger models (or even different sign patterns for this smaller model) and/or different categories of information, the conditions for resolving the analytic issues at stake can be more complicated. The purpose of this book is to contribute to an understanding of the circumstances under which less than fully quantitative, i.e., nonparametric, information can be used to resolve the issues of comparative statics and stability of mathematical models of phenomena.

1.5 ORGANIZATION OF THE BOOK

This book is divided into eight chapters. Chapters 2–4 examine what can be learned about the comparative statics and stability of systems defined by equations (1.1) through (1.7) by using only information about the entries of the Jacobian matrix and the *b* vector. Chapter 2 examines the classic qualitative case, where only sign pattern information is available to the researcher. The chapter contains necessary and sufficient conditions for full and partial sign solvability. Necessary and sufficient conditions are also given for sign stability. These results have been available for some time and set the stage for the extensions of the analysis presented in the remainder of the book. A mathematical appendix to the chapter reports results from matrix analysis and stability theory particularly useful in analyzing signed systems.

Chapter 3 extends the analysis to cover other cases for which sign pattern information is not sufficient, but which can nevertheless be solved nonparametrically. The strategy of analysis is to develop a hierarchy of information typologies for the entries of the Jacobian matrix, and then find conditions under which (some or all of) the elements of the inverse Jacobian matrix can be signed. Some of the procedures used are developed algorithmically. Chapter 4 contains examples.

Chapters 5–8 examine the same issues when the Jacobian matrix must be consistent with principles governing the system being studied. By this means particular quantitative matrices that are consistent with sign and other nonparametric restrictions, but are not consistent with the basic principles governing the system can be eliminated from consideration. Chapter 5 is our first demonstration of how augmenting information about the entries of the Jacobian matrix with information about the nature of the equilibrium system itself can lead to definitive results. In this chapter the assumption that the equilibrium equations arise from an optimization problem is shown to have pervasive implications for sign solvability. For an unconstrained maximization, as is the case with firms maximizing profit, the assumption of a regular maximum implies that the matrix A in (1.4) is negative definite. When the equations arise from constrained maximization, such as individual choice under a limited budget, the matrix A is negative definite under constraint. This additional information is shown to permit definite conclusions where sign information alone would not.

Chapter 6 explores how invoking the correspondence principle, a system-wide property, can lead to sign solvability. The correspondence principle hypothesizes that equilibrium is stable in the sense that the qualitative matrix DA in (1.7) has characteristic roots with real parts negative, and then uses this (quantitative) information to derive comparative statics results. The additional scope for unambiguous solutions of the qualitative system (1.4) under the correspondence principle is shown to be fairly limited.

Chapters 7 and 8 deal with how system-wide properties tying all the equations together coupled with qualitative information can lead to definitive results. In particular, the quantitative restrictions imposed on models of competitive economies by Walras's law and homogeneity are incorporated into the analysis of general equilibrium systems, in which the pattern of substitutes and complements is assumed to be known. Chapter 7 considers how these system-wide restrictions affect our understanding of comparative statics. Chapter 8 examines the implications for stability. The theme of this book is that definite conclusions about a model's predictions can be reached in many fields of scientific inquiry even though complete quantitative information is not available. In the least informed cases, sign information and other ordinal information can be sufficient for unambiguous predictions in nontrivial models. When they are not, the methods discussed in Chapter 3 identify how additional information might be used to resolve ambiguity. These insights have obvious implications for the design of data research programs.

In addition to information about specific components, scientific models are held together by underlying principles. Maximization, stability, and equilibrium can all impose additional restrictions that can resolve uncertainties relative to a purely qualitative environment. Other principles, taken from outside economies, may also lead to definitive results beyond those reported here. Our hope is that researchers in other fields will build upon this work to establish when a mathematical representation of reality makes unambiguous predictions, i.e., has scientific content.

2

Qualitative Comparative Statics and Stability

2.1 INTRODUCTION

We begin our analysis with a study of purely qualitative systems, i.e., systems in which the *only* information assumed to be available is sign pattern information concerning the entries of the matrix A and the vector b in (1.4). We first take up the problem of sign solvability. We then investigate sign stability. The appendix to this chapter contains the mathematical results that are necessary for the reader to follow the formal arguments.

2.2 SIGN SOLVABILITY—BACKGROUND

Definitions

A vector x determined by a system of linear equations $Ax = b$ is said to be *fully sign solvable* if for any matrix B with the sign pattern of A and vector c with the sign pattern of b,

$$By = c$$

implies that y has the sign pattern of x. For example,

$$\begin{pmatrix} - & + \\ - & - \end{pmatrix}\begin{pmatrix} x_1 \\ x_2 \end{pmatrix} = \begin{pmatrix} - \\ 0 \end{pmatrix}$$

is sign solvable with $x_1 > 0$, $x_2 < 0$.

A vector x determined by a system of linear equations as above is said to be *partially sign solvable* if there is a partition of $x = (x^1, x^2)$ where x^1 is sign solvable and x^2 is not. For example,

$$\begin{pmatrix} - & 0 \\ - & - \end{pmatrix}\begin{pmatrix} x_1 \\ x_2 \end{pmatrix} = \begin{pmatrix} - \\ - \end{pmatrix}$$

is sign solvable for $x_1 > 0$, but not for x_2.

Formally, for any scalar a, define $\operatorname{sgn} a = 0$ if and only if $a = 0$; $\operatorname{sgn} a = 1$ if and only if $a > 0$; and $\operatorname{sgn} a = -1$ if and only if $a < 0$. Given a real matrix $A = [a_{ij}]$ then $\operatorname{sgn} A = [\operatorname{sgn} a_{ij}]$. We now introduce the set $Q_A = \{B \mid \operatorname{sgn} B = \operatorname{sgn} A\}$. Thus, Q_A is the set of all matrices with the same sign pattern as A.

The issues of full and/or partial sign solvability arise in connection with systems of the form

$$Ax = b, \tag{2.1}$$

where A and b are specified only in terms of their sign patterns $(+, -, 0)$. The system (2.1) so specified is said to be fully *sign solvable* if and only if the sign of every entry of x is determined. Thus, for any matrix B in Q_A and any vector c such that $\operatorname{sgn} c = \operatorname{sgn} b$, the entries of the solutions y to $By = c$ will have the same sign as the corresponding entries in x. Partially signed systems are defined similarly.

Qualitative Economics

A traditional starting point for qualitative analysis is the discussion provided for economists in a brief initial section of Samuelson's *Foundations of Economic Analysis* (1947, 23–29) called "A Calculus of Qualitative Relations." In this section, Samuelson confronts the issue of whether or not a comparative statics analytic framework (i.e., an equation system's Jacobian matrix and b matrix) can yield definite results given only "a general feeling for the direction of things" (i.e., the signs of the matrix's entries). Then, as now, the question is compelling, since the quantification of the derivatives associated with a model's solution may not be possible on the basis of theoretical principles, and even if accomplished in practice, will often be transitory and error prone. Samuelson considered a few small examples to illustrate the analytic problem, and then despaired. The conclusion reached was that the chances that the conditions for a successful analysis would be satisfied were simply too small for the possibility to be taken seriously.

The specific example he cited concerned the computation of a single parameter sensitivity. Using Cramer's rule, this sensitivity can be expressed as the ratio of two determinants, that of the Jacobian matrix itself and that of the same matrix with the appropriate column replaced with the right-hand side of the linear system being manipulated. For $n \times n$ matrices, there are $n!$-many terms being summed to the value of each determinant. Accordingly, for (say) $n = 10$, each determinant will involve the sum of over three million terms. From Samuelson's perspective, "Regarded simply as a problem in probability, the chance that a run of this length should always have one sign is about one out of one with a million zeros after it." As a result, the issue is abandoned in favor of processing information about the Jacobian matrix in addition to the signs of its entries.

There is no doubt that the conditions for a purely qualitative analysis are restrictive, and to that degree "improbable." And further, it is clear, as Samuelson pointed out, that other information usually would be needed to resolve qualitative questions (e.g., the maximization hypothesis). Nevertheless, the issue at stake that motivates the consideration of a purely qualitative analysis remains important. It is unfortunately true that sometimes only the signs of sensitivities within a model's system of relationships can be safely assumed. Given this, a thorough review of how a qualitative analysis could be performed remains interesting, if only to have it in hand against a rare chance to use it. And besides, the situation is not so bad as Samuelson supposed. For a variety of reasons, opportunities for a successful qualitative analysis of applied models are not impossibly rare.[1]

The most important circumstance that serves to moderate the "improbability" of a successful qualitative analysis is the fact that the Jacobian matrices corresponding to applied models can have many zero entries. The conditions for a successful analysis are only that the nonzero terms in the expansions of determinants have the same sign, and there may be far fewer of these than $n!$ for an n-variable system. Further, as discussed in Chapter 3, even if a qualitative analysis fails, the number of "wrong" signed terms may be small. As a result, through qualitative analysis, researchers can identify precisely what extraqualitative information they would need to reach unambiguous conclusions. In addition, conditions necessary and sufficient for full or partial sign solvability in the purely qualitative case are still valid when quantitative information is available as well. Thus, resolving (or at least confronting) the purely qualitative problems of full and partial sign solvability is an essential first step in handling the more common analytical problem in which other as well as qualitative information is present.

2.3 THE ALGORITHMIC APPROACH TO STRONG SIGN SOLVABILITY

The Standard Form Algorithm

Lancaster (1962) published the first formal attempt to develop necessary and sufficient conditions for sign solvability. Lancaster conjectured that if a system could be shown to yield qualitative results, then it could also be shown to be manipulable into a particular, standard form. In essence, systems with qualitatively decidable attributes could be shown to "look" a certain way. This approach required the identification of "manipulations" for the linear system (2.1), that would not disturb, i.e., add to or detract from, its qualitative attributes. Specifically, for $S = [A; b]$ the augmented $n \times (n+1)$ matrix, allowable manipulations would include multiplying rows or columns of S by -1 and/or exchanging rows or columns of S. Because it will be useful in presenting Lancaster's elimination principle, discussed later in this section, we will develop Lancaster's standard form constructively by finding a qualitative solution and then developing conditions for the solution found to be the only qualitative solution.

To limit the system to be studied, assume that the matrix A is irreducible, i.e., assume that the entire system (2.1) must be solved simultaneously (see the appendix to this chapter). An analysis of reducible systems can be expressed in terms of the results found for irreducible systems applied to a structure of (antisymmetrically) interrelated irreducible systems. Further, reindex the columns of A as necessary to render all main diagonal entries nonzero. This manipulation is one of the allowable transformations identified by Lancaster and amounts to simply reindexing variables. The manipulation amounts to finding a nonzero term in det A's expansion and reindexing to bring the entries in this term onto the diagonal (see Theorem A2.2 of the appendix to this chapter).

Call the vector of signs, sgn x, a feasible sign pattern if and only if there exist magnitudes that could be assigned to the entries of sgn A and sgn b such that the solution vector to the system $Ax = b$ has the given sign pattern. For a given sgn A and sgn b, sgn x is a feasible sign pattern if and only if

(i) for the ith element of sgn b nonzero, there is at least one k such that in the ith row of the array $\text{sgn}[a_{ij}x_j]$, $\text{sgn}\, a_{ik}x_k = \text{sgn}\, b_i$; and
(ii) for the ith element of sgn b zero, the ith row of the array $\text{sgn}[a_{ij}x_j]$ has either both a positive and negative element or all zero entries.

The reasoning for both conditions is immediate. If sgn b_i is nonzero, then at least one of the terms being summed to this result must have the same sign. If sgn b_i is zero, then either there are both positive and negative terms with each group summing to the same absolute value or all terms are zero.

Given sgn A irreducible with all main diagonal entries nonzero and sgn b with some entry nonzero, a qualitative solution sgn x can be constructed as follows:

Step 1. The vector sgn b must have at least one nonzero sign. Select one of these and reindex rows of $S = [A; b]$ (an allowable transformation) such that it is the first element of sgn b, sgn $b_1 \neq 0$.
Step 2. Select sgn x_1 such that sgn $a_{11}x_1 = \text{sgn } b_1$.
Step 3. Consider the nonzeros in the first column of sgn A. For each k such that sgn $a_{k1} \neq 0$, select sgn x_k such that sgn $a_{kk}x_k = -\text{sgn } a_{k1}x_1$.
Step 4. Consider the nonzero entries in each of the columns, k, for which entries of sgn x were signed in Step 3. For each nonzero, repeat Step 3 for the entries of sgn x not signed. If in a given row, g, more than one element a_{gk} is nonzero and sgn x_g is unsigned, select the value of sgn x_g such that sgn $a_{gg}x_g = -\text{sgn } a_{gk}x_k$ for the smallest value of k such that $a_{gk} \neq 0$.
Step 5. Repeat Step 4 as long as entries of sgn x can be signed. Since A is assumed to be irreducible, the process will continue until all of the entries of sgn x have been given signs. Let sgn x^* denote the vector of signs constructed in this way.

Consider the array of signs sgn$[a_{ij}x_j^*]$. For the first row, sgn $a_{11}x_1^* = \text{sgn } b_1$. For the remaining rows, each row has both a positive and negative element. Accordingly, sgn x^* conforms to feasible sign pattern conditions (i) and (ii) and is a feasible sign pattern. For a sign solvable system, the issue becomes that of finding conditions that establish sgn x^* as the only feasible sign pattern.

To begin, note that the constructed solution is such that rows $2, 3, \ldots, n$ of sgn$[a_{ij}x_j^*]$ will have both a positive and a negative element. If sgn x^* is the only feasible sign pattern, then the first row must not have both a positive and negative element. If all rows contained both positive and negative entries, then $-\text{sgn } x^*$ would also be a feasible sign pattern.

Condition 1. If sgn x^* is the only feasible sign pattern to (2.1) for sgn $b_1 \neq 0$, then for sgn $a_{ij} \neq 0$, sgn $a_{1j}x_j^* = \text{sgn } b_1$.

Next, assume condition 1 is satisfied, and then consider the entries sgn b_i, $i \neq 1$. If one of these, say, the hth, is nonzero, then the hth and the 1st rows of S could be swapped (an allowable transformation) and the construction repeated, starting with the new row 1.[2] Call the result of this second construction sgn x^{**}. But now the hth row (formerly the 1st row) of sgn$[a_{ij}x_j^{**}]$ will contain both a positive and negative element. As a result, sgn $x^* \neq \text{sgn } x^{**}$, and the system is not sign solvable.

Condition 2. If $\operatorname{sgn} x^*$ is the only feasible sign pattern to (3.1) for $\operatorname{sgn} b_1 \neq 0$, then $\operatorname{sgn} b_i = 0$ for $i \neq 1$.

Finally, consider the subsystem

$$\sum_{j=2}^{n} a_{ij}x_j = -a_{i1}x_1, i = 2, 3, \ldots, n. \tag{2.2}$$

It is clear that if (2.1) is sign solvable, then (2.2) must be sign solvable as well. Assume that the submatrix $[a_{ij}]$, $i, j = 2, 3, \ldots, n$ is also irreducible. Given this, conditions 1 and 2 now can be applied to (2.2).

Condition 3. If (2.1) is sign solvable and $[a_{ij}]$ in (2.2) is irreducible, then there is only one nonzero element a_{i1}, $i \neq 1$ (from condition 2); and for this (setting $i = 2$), if $a_{2g} \neq 0$, then $\operatorname{sgn} a_{2g}x_g^* = -\operatorname{sgn} a_{21}x_1^*$ (from condition 1).

The analysis can continue to be applied to a sequence of subsystems, where the Mth subsystem is formed by the removal of the first $M - 1$ rows and columns of S. So long as each new subarray is irreducible, condition 3 can be invoked as a requirement for each subsystem. As a result, for (2.1) sign solvable, conditions 1–3 require that (for A reindexed as the conditions are applied) the matrix A have all zero entries below the first submain diagonal. This particular pattern of zero entries is called the (upper) Hessenberg form (Householder 1964). Further, if the signs of the system are changed such that $\operatorname{sgn} b_1$ and the entries of $\operatorname{sgn} x^*$ are all positive, then from conditions 1–3 it can be shown that the augmented matrix, S, of a sign solvable system can be manipulated so that it looks like this:

$$[\operatorname{sgn} A, \operatorname{sgn} b] = \begin{bmatrix} 1 & 1 & 1 & \cdots & 1 & 1 & 1 \\ -1 & 1 & 1 & \cdots & 1 & 1 & 0 \\ 0 & -1 & 1 & \cdots & 1 & 1 & 0 \\ 0 & 0 & -1 & \cdots & 1 & 1 & 0 \\ \cdot & \cdot & \cdot & \cdots & \cdot & \cdot & \cdot \\ \cdot & \cdot & \cdot & \cdots & \cdot & \cdot & \cdot \\ \cdot & \cdot & \cdot & \cdots & \cdot & \cdot & \cdot \\ 0 & 0 & 0 & \cdots & -1 & 1 & 0 \end{bmatrix}. \tag{2.3}$$

This is Lancaster's standard form.[3]

As it works out, the form (2.3) is sufficient but not necessary for sign solvability. The key assumption in the derivation of (2.3) is the irreducibility of each subsystem formed by deleting rows and columns of S. For A irreducible, (2.3) can be sign solvable without requiring that the various submatrices of A be irreducible. This circumstance was noticed by Gorman (1964), who proposed a sufficient algorithm for constructing sign solvable systems that did not require irreducible submatrices. Given this, Gorman's systems could not be put into the form (2.3). Gorman conjectured, but did not demonstrate, that his algorithm was also necessary.

Lancaster (1964) replied by noting that the augmented matrix S of any of Gorman's sign solvable systems could be partitioned into submatrices. If any of these possessed the sequence of irreducible substructures assumed above when applying condition 3, then the resulting submatrix would correspond to the original standard form. Unfortunately, this insight did not provide a revised, or equivalent, standard form, since the point at which a submatrix could not be further partitioned into its own submatrices, which in turn could be further partitioned, and so on, is not (or at least has not been shown to be) constrained by the property of sign solvability. As a result, the idea that a sign solvable system could be detected by its appearance was not sustained.

Lancaster (1964) did show that the matrix sgn A in (2.3) had the fewest zeros for the coefficient matrix of a sign solvable system. This result is an extremely convenient necessary condition for sign solvability to bring to applied contexts. For $Ax = b$ and A $n \times n$ irreducible, if there are M-many nonzeros in sgn A, then for

$$M > ((n^2 + 3n - 2)/2),$$

the system cannot be sign solvable. Lady (1983) applied a similar route of derivation by constructing a feasible solution sign pattern and then developing conditions such that the sign pattern was the only feasible sign pattern. Lady's conditions were necessary and sufficient. Klee, Ladner, and Manber (1984) showed that Lady's and Gorman's conditions were equivalent.

Finding Qualitative Solutions through Elimination

An alternative to detecting sign solvable systems by establishing their conformance to a particular form is to inspect all possible sgn patterns, identify which of these are feasible, i.e., satisfy (2.1) and (2.2), and for the feasible patterns determine if any solution variables have the same sign for all of them. This is Lancaster's (1966) elimination algorithm. The algorithm has the considerable advantage that the system being studied does not have to be manipulated into any particular form prior to using the algorithm. There are no standard forms to detect or contingent manipulations required during the algorithm's execution. In addition, distinctions between partial and full sign solvability do not impose new requirements for the analysis. Finally, testing a candidate qualitative solution for conformance with (2.1) and (2.2) is extremely simple to formulate as an algorithmic principle.

For these reasons, it is no surprise that the elimination algorithm has been a popular approach for investigating actual systems for qualitatively decidable characteristics (e.g., Ritschard 1983). Further, as discussed in Chapter 3, the elimination principle provides a ready frame of reference within which to introduce additional information to assist in resolving problems for which sign pattern information is not enough. In fact Gillen and Guccione (1990) showed how the principle can be used to find exactly those coefficients within the matrix S about which more than sign information is necessary in order to resolve the sign of a variable's sensitivity to a parameter change.

The major difficulty with the elimination principle is the need to construct and test all possible sign patterns. For A $n \times n$, if solution sign patterns with zeros are considered, the number of possible sign patterns is 3^n. If A is irreducible, then solutions with zero entries can be ignored, but this still leaves 2^n-many sign patterns to construct and test. These problems are moderated as computing technology advances, but they place inevitable constraints on the size of system that can be studied. Alternatively, the algebraic methods developed in the next section do not involve the same, inevitable increase in the computational burden of their application (although there are problems that become generally more acute as the system being studied becomes larger). Further, when the algebraic approach, with its graph-theoretic underpinning, is applied to actual systems, it provides greater insight about the interrelationships among a system's variables and parameters. Still, for sufficiently small systems the elimination principle is a ready resource for seeing if the system has any qualitatively decidable relationships.

2.4 THE ALGEBRAIC APPROACH TO SIGN SOLVABILITY

Algebraic methods of finding conditions for full or partial sign solvability derive conditions directly from the formulae used to compute a matrix determinant and adjoint.

Full Sign Solvability

The first proof of necessary and sufficient conditions for full sign solvability appears in Bassett, Maybee, and Quirk (BMQ) (1968). The BMQ theorem is not restricted to strong sign solvability, but instead gives necessary and sufficient conditions for signing both zero and nonzero entries of the x vector. The approach adopted in BMQ uses the concepts of *paths*, which are called chains in the older literature, and *cycles* in matrices together with Maybee's determinantal formulas (see the appendix to this chapter).

For the $n \times n$ matrix A, we call the product $A_p = a_{i(1)i(2)} a_{i(2)i(3)} \cdots a_{i(r-1)i(r)}$ a *path* in A if $r \geq 2$ and the set $V_p = \{i(1), i(2), \ldots, i(r-1)\}$ consists of distinct elements of N. The number $r-1$ is called the *length* of A_p and the set V_p is called the *index* set of A_p. If we wish to emphasize that $i(1) = j$ and $I(r) = k$ we will write $A_p[j \to k]$. Then j is the *initial index* of the path A_p, and k is the terminal index of the path.

The product $A_c = a_{i(1)i(2)} a_{i(2)i(3)} \cdots a_{i(r)i(1)}$ is called a *cycle* of A if $r \geq 2$ and the product $a_{i(1)i(2)} \cdots a_{i(r-1)i(r)}$ is a path in A. The *length* of the cycle is the number r, and the set $V_c = \{i(1), i(2), \ldots, i(r)\}$ is called the *index* set of the cycle. The diagonal entries $a_{i(q)i(q)}$ can also be considered cycles of A, and they have length one and index set consisting of the single index $V_c = \{i(q)\}$. If we wish to emphasize that a cycle A_c contains a particular index j, we write $A_c[j]$.

Maybee's (1966a) determinant formula plays a central role in the proofs of many of the propositions of qualitative economics. The formulas shows that negative cycles of length n enter into det A with sign $(-1)^n$, while negative cycles of length r enter principal minors of A of order r with sign $(-1)^r$. A particularly important implication of the formula is that all the nonzero terms in the expansion of det A are of sign $(-1)^n$ if and only if all cycles in A are nonpositive. BMQ also adopt a standard form of the qualitatively specified system but it is quite different from the standard form developed by Lancaster. The BMQ standard form of a sign solvable system has negative diagonal entries in the coefficient matrix, the solution vector is nonpositive, and entries in the right-hand vector are nonnegative. Since every fully sign solvable system can be transformed into this standard form, without loss of generality, the BMQ theorem specifies necessary and sufficient conditions for full sign solvability for systems already in this standard form.

THEOREM 2.1 (Bassett, Maybee, and Quirk 1968). *The system $Ax = b$ is fully sign solvable if and only if, by admissible operations, it can be* *transformed into the system $By = c$ where*

(1) *diagonal entries in B are negative, and $c \geq 0$;*
(2) *all cycles of B are nonpositive; and*
(3) *$c_k \neq 0$ implies that every path $b(i \to k)$ in B is nonnegative.*

PROOF. $B \in Q_A$ is nonsingular. Assume not. Then there exists $w \neq 0$ such that $Bw = 0$, which implies $B(tw) = 0$ for any scalar t. Since $By = c$ implies $B(tw + y) = c$ for any scalar t, clearly $w_k \neq 0$ implies the kth element of y is unsigned. Hence $Ax = b$ is not fully sign solvable.

Given that $B \in Q_A$ is nonsingular, renumber equations by premultiplying by a permutation matrix P that brings a nonzero term to the diagonal, as in $PAx = Pb$. Multiply certain variables by (-1) by premultiplying x by a diagonal matrix D, and postmultiplying PA by D^{-1}, where $d_{ii} = 1$ if $x_i \leq 0$, and $d_{ii} = -1$ if $x_i > 0$: $PAD^{-1}(Dx) = Pb$. Thus, the solution vector Dx consists of nonpositive terms. Finally, multiply certain equations by (-1) by premultiplying on both sides by the diagonal matrix M such that $m_{ii} = -1$ if the diagonal entry of PAD^{-1} is positive and by $m_{ii} = 1$ otherwise.

The resulting system can be written as $By = c$, where $B = MPAD^{-1}$, $y = Dx$, and $c = MPb$. To sign det B for every matrix with the same sign pattern of B, every term in the expansion of the determinant must be weakly of the same sign. The necessary and sufficient condition for this is that all cycles in B be nonpositive. In $y = B^{-1}c$, diagonal entries of B^{-1} are negative while the y vector is nonpositive. Hence under sign solvability, the c vector must be nonnegative, and every term in $B^{-1}c$ must be nonpositive, which implies that if $c_k > 0$, then every nonzero path $b(i \to k)$ must be positive.

The BMQ result has led to a number of later papers, dealing with a variety of questions concerning sign solvable systems and SNS matrices (sign nonsingular matrices, i.e., matrices with signed nonzero determinants), including the further specification of sign solvable systems, and the development of efficient algorithms for identifying sign solvable systems.

Manber (1982) showed that any fully sign solvable system can be decomposed into the system

$$\begin{bmatrix} A_1 & 0 \\ A_2 & A_3 \end{bmatrix} \begin{bmatrix} x^1 \\ x^2 \end{bmatrix} = \begin{bmatrix} b^1 \\ b^2 \end{bmatrix},$$

where $b^1 = 0$, $x^1 = 0$, and $A_3 x^2 = b^2$ is strongly sign solvable, i.e., where $x_i \in x^2$ implies $x_i \neq 0$. Furthermore, she proved that the set of vectors b for which $Ax = b$ (in standard form) is sign solvable is precisely that b

for which $b_i \leq 0$ for i such that $a(j \to i) \geq 0$ for all j, with $b_i = 0$ for all other i.

Yamada (1988) noted that implicit in the BMQ approach is the assumption that there are no cancellations possible from the numerators and denominators in solving for x. In general, problems arising from this can be avoided by dealing with irreducible matrices A. However, Yamada also showed that so long as $Ax = b$ is in standard form, no cancellations are possible, so that the BMQ conditions hold for any system $Ax = b$ in standard form, reducible or irreducible.

Lady (1983) directed his attention to the development of an algorithmic approach to sign solvability, reminiscent of the earlier Lancaster and Gorman approaches. The algorithm begins by determining a set of signs for x consistent with a solution to $Ax = b$ (steps 1–5 above), and then uses sequential elimination methods to obtain a fully sign solvable solution (if one exists). Lady deals with the case of strong solvability with an irreducible matrix A. In effect, Lady proves a revised version of the Gorman conjecture, i.e., that the Gorman algorithm is both necessary and sufficient for strong sign solvability, although this is not stated formally in the paper.

Lady also gave necessary and sufficient conditions for qualitative invertibility of a matrix, without transforming to the standard form of BMQ. He also establishes the important result that if A is irreducible and qualitatively invertible, and if we denote A^{-1} by $B = [b_{ij}]$, then $a_{ij} \neq 0$ implies that sgn b_{ji} = sgn a_{ij}. Lady and Maybee (1983) extended this to the case of A irreducible, with negative diagonal entries, and qualitatively invertible, by showing that if $a_{ij} = 0$, then b_{ji} is signed unambiguously if and only if every path $a(j \to i)$ in A has weakly the same sign, in which case sgn b_{ji} is opposite to that sign. The idea of an "essential zero" is introduced, namely, a zero entry that if changed to a nonzero entry would render the matrix no longer sign nonsingular. It then follows that if a_{ij} is an essential zero then b_{ji} is indeterminate in sign.

Maybee and Weiner (1987) extended the BMQ approach to global and generally nonlinear functions (L mappings), prove global univalence results, and extended the Lady-Maybee results concerning the signing of the inverse to these global functions as well.

Hansen (1983) developed a solution algorithm for identifying fully sign solvable systems. Klee, Ladner, and Manber (1984) provided a formal proof of the Gorman conjecture for strong sign solvability, and develop an alternative algorithm for recognizing sign solvable systems. Klee (1987) explicitly examined efficiency issues in identifying sign solvability, and developed an efficient algorithm for sign solvability. Klee identified an S-matrix as an augmented matrix (given $Ax = b$,

$S = [A; b]$) associated with a sign solvable system. He showed that given an S-matrix, reversing the sign of any nonzero entry in S will render the matrix not an S-matrix, i.e., not sign solvable. Klee identified S*-matrices as S-matrices associated with strong sign solvable systems, and shows that, given an $m \times (m+1)$ S*-matrix, there are at least $2m$ nonzero entries in S^* and no more than $m(m+3)/2$ nonzero entries.

Albin and Gottinger (1983) investigate the complexity problem associated with identifying an SNS-matrix, i.e., a sign nonsingular matrix, and show that any SNS-matrix has at least $(1/2)n(n-3)+1$ zeros. Carlson (1988) provides a graph-theoretic characterization of sign nonsingular matrices. Lundgren and Maybee (1984) show how to construct digraphs for upper Hessenberg L-matrices of the type used by Lancaster in his original standard form.

Partial Sign Solvability

On partial sign solvability, i.e., solving for the signs of some but not necessarily all of the entries of the solution vector x, there was an unpublished paper by Quirk (1972a) that provided some introductory material and formed the basis for the treatment in Maybee (1981). Lady and Maybee (1983) provide an example of a matrix that is sign nonsingular but has a sign-indeterminant entry in every column of its inverse (and thus could not possibly be part of a fully sign solvable system). As it turns out, there can be partial sign solvability in a system in which the coefficient matrix A has signed nonzero determinant; or has signed zero determinant, i.e., $B \in Q_A$ implies that sgn det $B =$ sgn det $A = 0$; or has unsigned determinant, i.e., there exist $B, C \in Q_A$ such that sgn det $B \neq$ sgn det C. Examples covering these three cases are the following:

EXAMPLE 1. Here, A has signed nonzero determinant:

$$\begin{bmatrix} - & + & 0 \\ - & - & + \\ 0 & - & - \end{bmatrix} \begin{bmatrix} x_1 \\ x_2 \\ x_3 \end{bmatrix} = \begin{bmatrix} - \\ - \\ 0 \end{bmatrix}.$$

Inverting, we obtain

$$\begin{bmatrix} x_1 \\ x_2 \\ x_3 \end{bmatrix} = \begin{bmatrix} - & - & - \\ + & - & - \\ - & + & - \end{bmatrix} \begin{bmatrix} - \\ - \\ 0 \end{bmatrix} = \begin{bmatrix} + \\ ? \\ ? \end{bmatrix},$$

so that $x_1 > 0$, but x_2 and x_3 are unsigned.

EXAMPLE 2. Here, A has signed zero determinant, but $x_1 > 0$ from either equation 2 or 3, while x_2 and x_3 are unsigned:

$$\begin{bmatrix} - & + & + \\ - & 0 & 0 \\ + & 0 & 0 \end{bmatrix} \begin{bmatrix} x_1 \\ x_2 \\ x_3 \end{bmatrix} = \begin{bmatrix} - \\ - \\ + \end{bmatrix}.$$

EXAMPLE 3. Here, A has an unsigned determinant, but $x_1 > 0$, while x_2 and x_3 are unsigned:

$$\begin{bmatrix} - & 0 & 0 \\ + & - & + \\ - & + & - \end{bmatrix} \begin{bmatrix} x_1 \\ x_2 \\ x_3 \end{bmatrix} = \begin{bmatrix} - \\ 0 \\ 0 \end{bmatrix}.$$

The following theorem gives necessary and sufficient conditions for partial sign solvability for the general case.

THEOREM 2.2 (Quirk 1992). *Partition x into $x = (x^1, x^2)$, where the entries of x^1 are signed and those of x^2 are unsigned. Then $Ax = b$ is partially sign solvable for x^1 if and only if, by admissible operations, $Ax = b$ can be transformed into $By = c$, which appears as*

$$\begin{bmatrix} B_1 & 0 \\ B_2 & B_3 \end{bmatrix} \begin{bmatrix} y^* \\ y^{**} \end{bmatrix} = \begin{bmatrix} c^* \\ c^{**} \end{bmatrix},$$

where B_1 and B_3 are square blocks and 0 is a block of zeros. When y is partitioned as x, $y = (y^1, y^2)$, then y^1 is a subset of y^*; and $B_1 y^* = c^*$ satisfies the following conditions:

(1) *diagonal entries in B_1 are negative and $c^* \geq 0$;*
(2) *all cycles in B_1 are nonpositive; and*
(3) *$c_k^* \neq 0$, $y_i \in y^1$ implies that all nonzero paths $b(i \to k)$ in B_1 are positive.*

PROOF. If $B \in Q_A$ implies that B is nonsingular, then, letting $B = B_1$, conditions (1) through (3) are necessary and sufficient for signing x^1 in $Ax = b$, from the argument of the previous theorem. Thus, the only cases of interest are those in which $B \in Q_A$ implies that $\det B = 0$ (A has signed zero determinant) or in which there exist $B, C \in Q_A$ such that $\operatorname{sgn} \det B \neq \operatorname{sgn} \det C$ (A has unsigned determinant). We consider each case in turn.

When $B \in Q_A$ implies $\det B = 0$, then for any $B \in Q_A$, there exists $w \neq 0$ such that $Bw = 0$, which implies $B(tw) = 0$ for any scalar t. It follows that if $Ax = b$ is sign solvable for x^1 and w is partitioned into $w = (w^1, w^2)$, then $w^1 = 0$. Thus in $Aw = 0$, w^1 is sign solvable as $w^1 = 0$. Let $w = (w^*, w^{**})$, where w^* is the maximal subset of w such that, for every $B \in Q_A$, $Bw = 0$ implies $w^* = 0$. w^* is nonempty; in particular, $w^1 \subseteq w^*$.

Reindex $Aw = 0$ into

$$\begin{bmatrix} A_1 & A_4 \\ A_2 & A_3 \end{bmatrix} \begin{bmatrix} w^* \\ w^{**} \end{bmatrix} = \begin{bmatrix} 0 \\ 0 \end{bmatrix}.$$

Since $w^* = 0$, $A_3 w^{**} = 0$. Also, by hypothesis, entries of w^{**} are unsigned, so that, given any element w_k of w^{**}, there exists $B \in Q_A$ such that $Bw = 0$ implies that $w_k \neq 0$. Let $w^* = (w_1, \ldots, w_m)$, $w^{**} = (w_{m+1}, \ldots, w_n)$. Suppose that $a_{1,m+1} \neq 0$, and choose $B \in Q_A$ such that $B_3 w^{**} = 0$ implies that $w_{m+1} \neq 0$. Since $B_1 w^* = (-)B_4 w^{**}$, $D \in Q_A$ can be chosen so that $D_3 = B_3$, $|d_{1,m+1}| = M$, $|d_{ij}| \leq \varepsilon$ for $j = m+2, \ldots, n$. Clearly, for M sufficiently large and ε sufficiently small, $B_4 w^{**} \neq 0$, which contradicts w^* signed as $w^* = 0$. Since this argument applies to every entry in B_4, sign solvability of w^* as $w^* = 0$ implies that B_4 is a block of zeros, and the theorem holds. (In this case, the index set of w^* is that of x^1.)

Next consider the case in which A has unsigned determinant in the system $Ax = b$. First take the case where $b = 0$. $Ax = 0$ sign solvable for x^1 implies $x^1 = 0$. As before, let $x = (x^*, x^{**})$, where x^* is the maximal subset of x such that $Bx = 0$ implies $x^* = 0$ and $B \in Q_A$. By the argument of the previous section, $Ax = 0$ is reducible to the pattern shown in the statement of the theorem.

When $b \neq 0$, without loss of generality, assume that $Ax = b$ is transformed by admissible operations into the form where diagonal entries are negative, and with $b \geq 0$. Note that the assumption that A has unsigned determinant implies that A has at least two nonzero terms in the expansion of $\det A$, of opposite signs. Thus, by admissible operations, $Ax = b$ can be transformed into condition (1) of the theorem.

Let $x_i \in x^1$. Then, for all $B \in Q_A$ such that $\det B \neq 0$ and for all $c \in Q_b$,

$$x_i = \left\{ \sum_{J=1}^{m} B_{ji} c_j \right\} / (\det B),$$

where B_{ji} is the cofactor of the element b_{ji} in B. Hence, for all $c_j > 0$, partial sign solvability implies that $B_{ji}/\det B \leq 0$ for all $B \in Q_A$ for which $\det B \neq 0$.

Given $x_i \in x^1$ and given $c_k > 0$, it is first shown that every element in any positive cycle in A must appear in A_{ki}, the cofactor of a_{ki}.

Suppose that a_{rs} belongs to a positive cycle in A but a_{rs} does not appear in A_{ki} (so that $r = k$ or $s = i$). Choose $B \in Q_A$ with $b_{jj} = -2$ for all j, assign every element in the positive cycle containing b_{rs} an absolute value of 1, and choose all other nonzero entries in B arbitrarily small in absolute value. Then it is easy to verify that sgn $\det B = (-1)^n$, and sgn $B_{ki} = (-1)^{n+1}$.

But there exists $C \in Q_A$ with all entries in C as in B except that c_{rs} is chosen arbitrarily large in absolute value, so that sgn $\det C = (-1)^{n+1}$, since the product of positive cycles and negative diagonal entries enter $\det C$ with sgn$(-1)^{n+1}$. However, c_{rs} does not appear in C_{ki}, $C_{ki} = B_{ki}$; thus sgn $C_{ki}/\det C \neq$ sgn $B_{ki}/\det B$. This violates the sign solvability conditions for x_i.

Consider now the problem of signing $B_{ki}/\det B$, given that all entries in all positive cycles in B appear in B_{ki}. If A is irreducible, then it is possible to choose $C \in Q_A$ such that, simultaneously, $C_{ki} = 0$ and $\det C \neq 0$, since there exist terms in the expansion of $\det C$ that do not appear in C_{ki}. By continuity, small variations in the values assigned to the entries of the positive cycles will change the sign of c_{ki} from zero to positive and from zero to negative, while preserving the sign of the determinant, as well as preserving the sign pattern. By successive application of this argument, we obtain the conditions of the theorem.

2.5 SIGN STABILITY

A real $n \times n$ matrix A is said to be *sign stable* if, given any $B \in Q_A$, B is a stable matrix, i.e., the real parts of the eigenvalues (characteristic roots) of B are negative. Also, A is said to be potentially stable if there exists a matrix $B \in Q_A$ that is a stable matrix. Finally, A is called a sign semistable matrix if, given any $B \in Q_A$, B is semistable, i.e., the real parts of the eigenvalues of B are nonpositive. Recall that the linear approximation to the dynamic adjustment process $dy/dt = g(f)$ discussed in Chapter 1 (1.7) is

$$dy/dt = DA(y - y^*),$$

where D is a diagonal matrix with positive diagonal.

Samuelson (1947, 438) indicated the role that qualitatively specified matrices might play in stability analysis in economics: "Formally, our stability criteria are complete. But for most theoretical purposes numerical values of the various coefficients are not at hand. It would be highly desirable to be able to infer from the qualitative properties of our dynamic matrices even before their characteristic determinants have been (laboriously) expanded out whether or not they are stable. Unfortunately, this is rarely possible."

Matrices with Negative Main Diagonals

The first work done on investigating criteria for sign stability of matrices was by Quirk and Ruppert (1964), using an approach that predated Maybee's cyclic analysis of determinants. However, since the concepts of cycles of matrices are central to sign stability, the original Quirk-Ruppert conditions are restated in terms of cycles. As noted above, a matrix A is said to be sign stable if $B \in Q_A$ implies B is a stable matrix. That is, A is sign stable if the real parts of all of the characteristic roots of B are negative.

The analysis begins with the case where A has all diagonal entries negative.

THEOREM 2.3 (Quirk and Ruppert 1965). *Let A be an $n \times n$ real matrix with all diagonal entries negative. The A is sign stable if and only if*

(i) *every cycle in A of length two is nonpositive; and*
(ii) *every cycle in A of length greater than two is zero.*

PROOF. *Necessity.* By the Routh-Hurwitz conditions, if A is stable then $k_i > 0 \;\; i = 1, \dots, n$, where $k_i = (-1)^i x$ sum of all ith order principal minors of B, for $B \in Q_A$. Thus $k_1 > 0$ implies that all diagonal entries of A are nonpositive, with at least one negative, and $k_2 > 0$ implies

$$\sum_{j=1}^{n} \sum_{k>j} (b_{jj}b_{kk} - b_{jk}b_{kj}) > 0 \text{ for } B \in Q_A.$$

Hence if A is sign stable, $a_{ij}a_{ji} \leq 0$ for all $i \neq j$; i.e., condition (i) of the theorem must hold.

To show that sign stability of A implies that all cycles of A of length greater than two are zero, consider a matrix with a nonzero upper diagonal ($a_{i,i+1} \neq 0 \;\; i = 1, \dots, n-1$), with $a_{n1} \neq 0$, and with all other off-diagonal entries in A zero. Thus the matrix contains a nonzero cycle of length n, namely $a_{12}a_{23} \cdots a_{n-1,n}a_{n1}$. Since cycles of length n enter into k_n with coefficient $(-1)^{n+1}$, sign stability of A implies that $\operatorname{sgn} a_{12}a_{23} \cdots a_{n-1,n}a_{n1} = (-1)^{n+1}$ or zero. Assume that n is odd (a similar argument applies if n is even). One of the Routh-Hurwitz determinantal conditions is

$$\begin{vmatrix} k_1 & k_3 & \cdots & k_n & 0 \\ 1 & k_2 & \cdots & k_{n-1} & 0 \\ 0 & k_1 & \cdots & k_{n-2} & k_n \\ 0 & 1 & & \vdots & \vdots \\ \vdots & \vdots & & \vdots & \vdots \\ 0 & 0 & & \vdots & \vdots \end{vmatrix} > 0.$$

In the expansion of this determinant, k_n^2 appears only once, with coefficient $(-1)^{n+4}$ times a determinant that is positive by an earlier Routh-Hurwitz determinantal condition. Thus $(-1)^{n+4}(-1)^{n+1}a_{12}^2 a_{23}^2 \cdots a_{n1}^2 \geq 0$ if A is sign stable, hence sign stability implies $a_{12}a_{23} \cdots a_{n1} = 0$.

Clearly, replacing the zero entries in A with nonzero entries does not change this conclusion, since any nonzero cycle of length n will still appear in k_n as above. Moreover, this argument establishes condition (ii) as a necessary condition for sign stability, since every principal submatrix of a sign stable matrix with negative diagonal entries must also be sign stable.

Sufficiency. In establishing sufficiency of (i) and (ii) it is convenient to deal with the case where A is irreducible since A is stable if and only if every irreducible diagonal block of A is stable. Irreducibility combined with condition (ii) implies that A is combinatorially symmetric, i.e., $a_{ij} \neq 0$ implies $a_{ji} \neq 0$. Assume that this is not the case, i.e., that A is irreducible and there exists $a_{ij} \neq 0$ for which $a_{ji} = 0$. Reindex into $a_{12} \neq 0$ with $a_{21} = 0$. If row 2 has no nonzero entries to the right of the diagonal, then A is reducible. Hence reindex some nonzero element in row 2 into a_{23}. Then condition (ii) implies that $a_{31} = 0$. Continue with such reindexing until a row m is reached such that no nonzero entries exist to the right of the diagonal. If $m = n$, then every off-diagonal element in the first column is zero and A is reducible. If $m < n$, then the mth row has zeros to the right of the diagonal, the first column has zeros in rows 2 through m, and A is reducible unless there exists a nonzero entry to the right of column m in rows 2 through m. Reindex the column index of any such element into $m + 1$, and continue, by setting $a_{m+1,1} \neq 0$. A continuation of this process leads to the desired result.

It is easy to show that if A is irreducible, then condition (ii) implies that there are exactly $n-1$ nonzero entries above the diagonal with $n-1$ symmetrically placed nonzero entries below the diagonal. Choose a diagonal matrix D with $d_{11} \neq 0$, and given $a_{ij} \neq 0$, let $d_{ii} = -(a_{ji}/a_{ij})d_{jj}$, $i, j = 2, \ldots, n$. This choice of D is consistent, since $a_{ij} \neq 0$, $a_{jk} \neq 0$ implies $a_{ki} = 0$, while condition (i) implies $d_{ii} > 0$ $i = 1, \ldots, n$. Consider $DA + A'D = [d_{ii}a_{ij} + d_{jj}a_{ji}] = C$. Then C is a diagonal matrix, with $c_{ii} = 2d_{ii}a_{ii}$ $i = 1, \ldots, n$, so that C is negative definite. Thus the Lyapunov criterion for stability is satisfied. Since this holds for every $B \in Q_A$, (i) and (ii) are necessary and sufficient for sign stability of A, given that all diagonal entries of A are negative.

To illustrate, consider the qualitative matrix Q_A, where the sign pattern of A is given by

$$\operatorname{sgn} A = \begin{bmatrix} - & + & 0 & 0 \\ - & - & + & 0 \\ 0 & - & - & + \\ 0 & 0 & - & - \end{bmatrix}.$$

In this example, all cycles of length two are nonpositive, and all cycles of length three or four are zero. It may be verified that every noncancelled term in the Routh-Hurwitz determinantal conditions for stability (see Gantmacher 1959) is positive, and that all other Routh-Hurwitz conditions for stability are also satisfied.

The sign stability and sign solvability results can be illustrated by a phase diagram representation of the dynamic adjustment of prices in a two-market partial equilibrium system. Let p_1, p_2 denote the prices of the two goods, and let α denote a shift parameter. $E_1 = E_1(p_1, p_2, \alpha)$ and $E_2 = E_2(p_1, p_2)$ are excess demands (demand minus supply) for the two goods, where α is assumed to shift E_1 directly only. Given $\alpha = \alpha^0$, an equilibrium of the system is a price vector $p^* = (p_1^*, p_2^*)$ such that

$$E_1(p_1^*, p_2^*, \alpha^0) = 0,$$

and

$$E_2(p_1^*, p_2^*) = 0.$$

Suppose now that, if the system is not in equilibrium, prices adjust according to the rules that price goes up if excess demand is positive and price goes down if excess demand is negative, i.e.,

$$\dot{p}_1 = g_1(E_1),$$

and

$$\dot{p}_2 = g_2(E_2)$$

where $g_1' > 0$, $g_2' > 0$, and where $g_1(0) = g_2(0) = 0$. Suppose that $\partial E_1/\partial p_1 < 0$, $\partial E_1/\partial p_2 > 0$, $\partial E_1/\partial a > 0$, $\partial E_2/\partial p_1 < 0$, $\partial E_2/\partial p_2 < 0$, so that, in linear approximation form, we have

$$\dot{p} = DA(p - p^*),$$

where

$$D = \begin{bmatrix} g_1' & 0 \\ 0 & g_2' \end{bmatrix},$$

and

$$A = \begin{bmatrix} \partial E_1/\partial p_1 & \partial E_1/\partial p_2 \\ \partial E_2/\partial p_1 & \partial E_2/\partial p_2 \end{bmatrix},$$

with

$$\text{sgn } DA = \begin{bmatrix} - & + \\ - & - \end{bmatrix},$$

so that A is sign stable. The phase diagram associated with the specification of A is shown in Figure 2.1. Note that all arrows point in towards the equilibrium, so that there is convergence over time to (p_1^*, p_2^*) for an arbitrary initial price vector, reflecting sign stability.

There also is full sign solvability for a new equilibrium, given a small change in α:

$$\begin{bmatrix} \partial E_1/\partial p_1 & \partial E_2/\partial p_2 \\ \partial E_2/\partial p_1 & \partial E_2/\partial p_2 \end{bmatrix} \begin{bmatrix} dp_1^*/d\alpha \\ dp_2^*/d\alpha \end{bmatrix} = \begin{bmatrix} -\partial E_1/\partial \alpha \\ 0 \end{bmatrix}$$

$$\Rightarrow \begin{bmatrix} - & + \\ - & - \end{bmatrix} \begin{bmatrix} dp_1^*/d\alpha \\ dp_2^*/d\alpha \end{bmatrix} = \begin{bmatrix} - \\ 0 \end{bmatrix}$$

$$\Rightarrow \begin{bmatrix} dp_1^*/d\alpha \\ dp_2^*/d\alpha \end{bmatrix} = \begin{bmatrix} - & - \\ + & - \end{bmatrix} \begin{bmatrix} - \\ 0 \end{bmatrix} = \begin{bmatrix} + \\ - \end{bmatrix}.$$

In terms of the phase diagram, the increase in α is represented by the new level curve $E_1' = 0$ and produces the outcome shown in Figure 2.2. Sign stability guarantees that the time path of prices will converge to the new equilibrium (p_1^{**}, p_2^{**}).

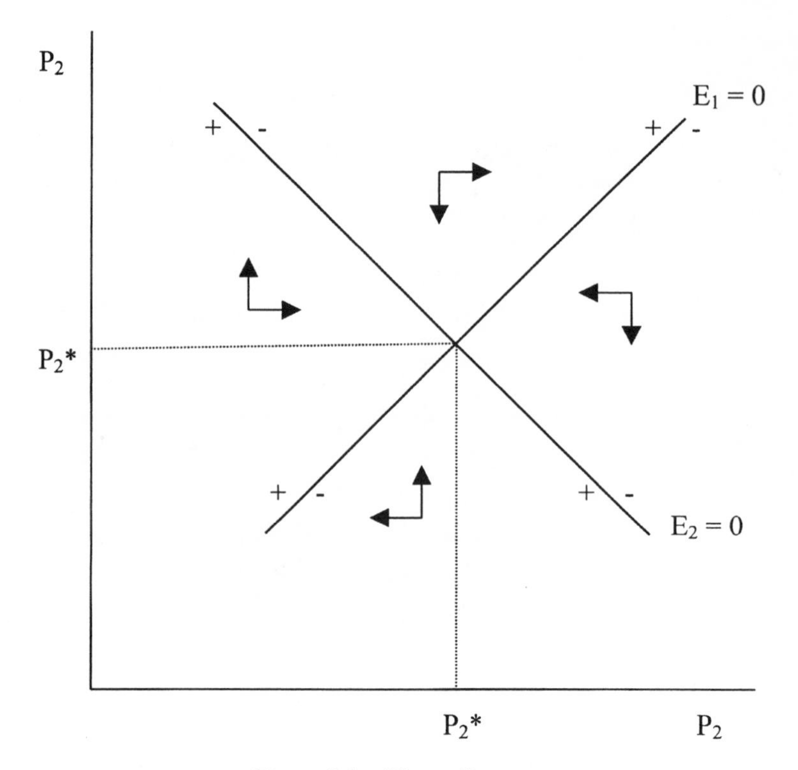

Figure 2.1. Phase diagram.

The General Case

Quirk and Ruppert (1965) attempted to extend their result to the case of matrices containing zero or positive diagonal entries. Their conditions for sign stability to cover this case are conditions (i) and (ii) above, plus two additional conditions: (iii) all diagonal entries are nonpositive with at least one diagonal element negative, and (iv) there exists a nonzero term in the expansion of det A. While these conditions are necessary for sign stability, they are not sufficient when A has zero diagonal entries, as shown by Jeffries (1974). The Jeffries counterexample to the general Quirk-Ruppert conditions is the following matrix:

$$A = \begin{bmatrix} 0 & -1 & 0 & 0 & 0 \\ 1 & 0 & -1 & 0 & 0 \\ 0 & 1 & -1 & -1 & 0 \\ 0 & 0 & 1 & 0 & -1 \\ 0 & 0 & 0 & 1 & 0 \end{bmatrix}.$$

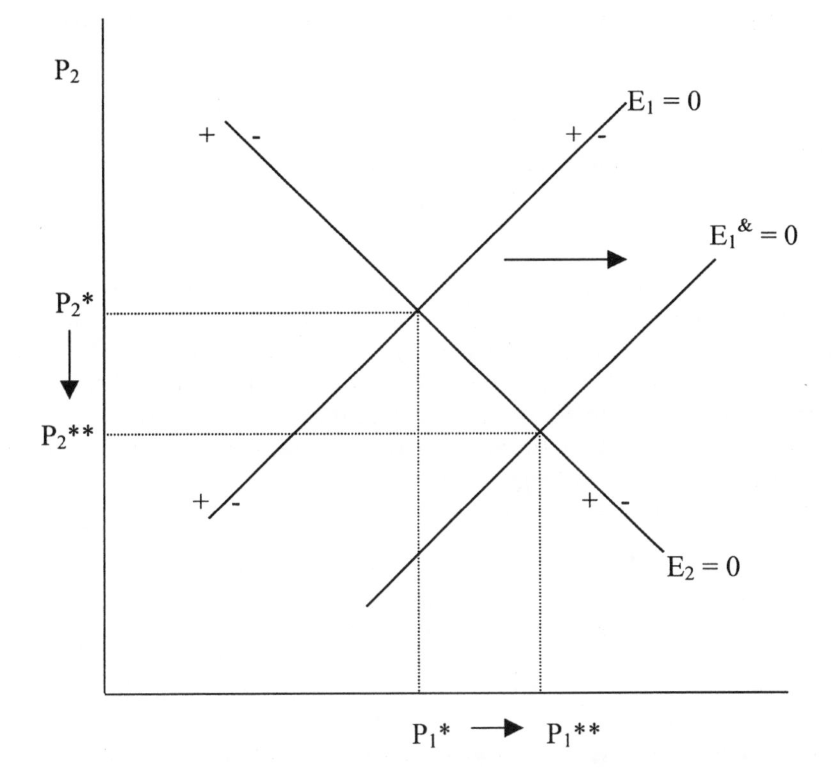

Figure 2.2. Increase in α.

This matrix satisfies conditions (i)–(iv) of Quirk-Ruppert, but two eigenvalues of A are $+i$ and $-i$, with zero real parts, so A is not stable.

The Jeffries conditions for sign stability are as follows.

THEOREM 2.4 (Jeffries, Klee, and van den Driessche 1977). *Let A be a real $n \times n$ matrix. Then A is sign stable if and only if*

(1) *all diagonal entries of A are nonpositive, with at least one diagonal element negative;*
(2) *all cycles in A of length two are nonpositive;*
(3) *all cycles in A of length three or more are zero;*
(4) *for every partition of N into sets B and W such that*
    (a) $a_{ii} \neq 0$ *implies* $i \in B$,
    (b) $a_{ii} \neq 0$ *implies there does not exist a unique* $j \in W$ *for which both* $a_{ij} \neq 0$ *and* $a_{ji} \neq 0$, *and*
    (c) $i \in W$ *implies there exists at least one* $j \in W$ *for which both* $a_{ij} \neq 0$ *and* $a_{ji} \neq 0$, *we have* $B = N$; *and*
(5) *some term in the expansion of* $\det A$ *is different from zero.*

We present an example to illustrate Theorem 2.4. Let A_1 have the sign pattern

$$\text{sgn } A_1 = \begin{bmatrix} - & - & 0 & 0 & 0 \\ + & 0 & - & 0 & 0 \\ 0 & + & 0 & - & + \\ 0 & 0 & + & 0 & 0 \\ 0 & 0 & - & 0 & 0 \end{bmatrix}.$$

Clearly conditions (1), (2), and (3) are satisfied. Suppose $1 \in B$ and $2 \notin B$. Then 2 is the unique integer in W such that a_{12} and a_{21} are not zero. Hence a partition satisfying (4) must be such that $2 \in B$ or else (4)(b) is violated. For the same reason we must also have $3 \in B$. But if either 4 or 5 belong to W condition (4)(c) is violated; hence $B = N$. Observe, however, that there is no nonzero term in the expansion of det A_1; hence A_1 is not sign stable.

Suppose we replace A_1 by A_2:

$$\text{sgn } A_2 = \begin{bmatrix} 0 & - & 0 & 0 & 0 \\ + & 0 & - & 0 & 0 \\ 0 & + & 0 & - & + \\ 0 & 0 & + & - & 0 \\ 0 & 0 & - & 0 & 0 \end{bmatrix}.$$

Then we have det $A_2 = a_{44}a_{35}a_{53}a_{21}a_{12} \neq 0$. Also if $4 \in B$, condition (iv) is satisfied and A_2 is sign stable.

Maybee (1989) has proved a theorem for sign semi-stable matrices. A matrix A is sign semi-stable if $B \in Q_A$ implies that the real part of every characteristic root of B is nonnegative or zero. To explain this result, we write the matrix A in the form $A = A_d + \tilde{A}$ where $A_d = \text{diag}[a_{11}, a_{22}, a_{nn}]$ and $\tilde{A} = A - A_d$. Let $u = (u_1, u_2, \ldots, u_n)$ be a complex vector. We say that u is g-orthogonal to A_d if $a_{ii} \neq 0$ implies $u_i = 0$. Thus if u is g-orthogonal to A_d, it is g-orthogonal to B_d for every matrix $B \in Q(A)$. We now work in the class of sign semi-stable matrices.

The following three theorems are valid for sign semi-stable matrices.

THEOREM 2.5. *The following are equivalent statements*:

1. *The matrix A is semistable.*
2. *The matrix A satisfies conditions* (1), (2), *and* (3) *of Theorem* 2.4.
3. *There exists a diagonal matrix* $D = \text{diag}[d_1, \ldots, d_n]$ *with* $d_i > 0$ *for* $i = 1, 2, \ldots, n$ *such that* $DAD^{-1} = A_d + \tilde{S}$, *where* $\tilde{S}$ *is a matrix in Frobenius normal form with skew symmetric diagonal submatrices and satisfies conditions* (2) *and* (3) *of Theorem* 2.4.

THEOREM 2.6. *The following are equivalent statements about a sign semistable matrix*:

1′. *Matrix* A *has* $\lambda = 0$ *as an eigenvalue.*
2′. *Every matrix in* Q_A *has* $\lambda = 0$ *as an eigenvalue.*
3′. *There is an eigenvector* u *satisfying* $Au = 0$ *that is g-orthogonal to* A_d.

THEOREM 2.7. *The following are equivalent statements about a sign semistable matrix*:

1″. *The matrix* A *does not have a purely imaginary* (*nonzero*) *eigenvalue.*
2″. *No matrix in* Q_A *has a purely imaginary eigenvalue.*
3″. *There is no eigenvector* u *satisfying* $Au = i\mu u$, $\mu \neq 0$ *that is g-orthogonal to* A_d.

Combining these results we obtain the following different version of the sign stability theorem.

THEOREM 2.8. *The real matrix* A *is sign stable if and only if*

(1) *the matrix* A *satisfies conditions* (*i*) *and* (*ii*) *of Theorem* 2.3;
(2) *all diagonal entries are nonpositive, with at least one diagonal element negative*; *and*
(3) *the matrix* A *does not have an eigenvector that is g-orthogonal to* A_d.

To illustrate these results consider the matrix

$$A = \begin{bmatrix} -1 & 1 & 0 & 0 & 0 \\ -1 & 0 & -1 & 0 & 0 \\ 0 & 1 & 0 & -1 & 1 \\ 0 & 0 & 1 & 0 & 0 \\ 0 & 0 & -1 & 0 & 0 \end{bmatrix}.$$

Any vector $u = \alpha(0,0,0,1,1)$ is an eigenvector g-orthogonal to A_d satisfying $Au = 0$. On the other hand, there is no complex vector satisfying $Au = i\mu u$ for $\mu \neq 0$.

For the Jeffries example above, we have $d(A) = -1$, so $\lambda = 0$ is not an eigenvalue. However, the vector $u = \alpha(1, i, 0, i, -1)$, which is g-orthogonal to A_d, is an eigenvector of A belonging to the eigenvalue $-i$.

The matrix such that

$$\operatorname{sgn} A = \begin{bmatrix} 0 & + & 0 & 0 & 0 \\ - & 0 & - & 0 & 0 \\ 0 & + & 0 & - & + \\ 0 & 0 & + & - & 0 \\ 0 & 0 & - & 0 & 0 \end{bmatrix}.$$

is a sign stable matrix, since it has no eigenvector g-orthogonal to A_d.

From the proof of Theorem 2.4 or from the proofs of Theorem 2.6 and 2.7, it is rather simple to derive linear algorithms for testing the sign stability of a matrix.

2.6 POTENTIAL STABILITY

The first systematic work on potential stability was done by Quirk (1968b). These results are also reported in Maybee and Quirk (1969). Other results related to this problem appear in Bone, Jeffries, and Klee (1988) and Jeffries and Johnson (1988). The general results appear in a paper of Johnson and van den Driessche (1988).

The basic approach is to consider a nested sequence of principal minors. To review the ideas let $\alpha_1, \alpha_2, \ldots, \alpha_n$ be a rearrangement of the set $N = \{1, 2, \ldots, n\}$, and $B[\{\alpha_1, \ldots, \alpha_k\}]$ denote the principal submatrix in rows and columns $\alpha_1, \ldots, \alpha_k$ of the $n \times n$ real matrix B. We say that sign pattern A allows a *nested sequence of properly signed principal minors* (abbreviated to a *properly signed nest*) if there exists $B \in Q_A$ and $\alpha_1, \alpha_2, \ldots, \alpha_n$ so that sign (det $B[\{\alpha_1, \ldots, \alpha_k\}] = (-1)^k$ for $k = 1, \ldots, n$.

Observe that this is equivalent to the existence of $C \in Q_A$ and a permutation matrix P such that $B = P^T C P$ has a *leading properly signed nest*, i.e.,

$$\operatorname{sign}(\det B[\{1, \ldots, k\}]) = (-1)^k \text{ for } k = 1, \ldots, n.$$

It is important to understand that this means that there is an actual matrix C with a properly signed nest; a nested sequence of index sets, each of which individually allows a properly signed minor, may not be sufficient.

Here is an example to illustrate this fact. Consider the sign pattern

$$A = \begin{bmatrix} - & + & 0 \\ - & + & + \\ 0 & + & + \end{bmatrix}.$$

A appears to allow a sequence of leading principal minors with proper signs. But such a sequence is not realizable by numbers, as can be seen by considering, without loss of generality,

$$C = \begin{bmatrix} -a & 1 & 0 \\ -d & b & 1 \\ 0 & e & c \end{bmatrix} \in Q_h,$$

with a, b, c, d, e all positive. A properly signed nest requires

$$-a < 0, d - ab > 0, c(d - ab) + ae < 0,$$

which is not possible. Note that A is not potentially stable.

The basic result connecting properly signed nests and potential stability is the following.

THEOREM 2.9. *If A has a sign pattern that allows a properly signed nest, then A is potentially stable. Moreover, A contains a nested sequence of potentially stable sign patterns of orders* $1, 2, \ldots, n$.

The proof of Theorem 2.9 rests upon the Fuller-Fisher theorem (Fuller and Fisher 1958), shown in the appendix to this chapter as A2.8. To illustrate Theorem 2.9, consider the tree sign pattern

$$A = \begin{bmatrix} - & + & 0 \\ - & - & + \\ 0 & - & + \end{bmatrix}.$$

Then

$$B = \begin{bmatrix} -1 & 1 & 0 \\ -1 & -1 & 1 \\ 0 & -3 & 1 \end{bmatrix} \in Q_A$$

has a leading properly signed nest, and the leading principal submatrices of B form a nested sequence of stable matrices. Note that $A[\{1, 2\}]$ is in fact sign stable, but A is not.

To identify classes of potentially stable matrices using Theorem 2.9, we need to know which sign patterns allow a properly signed nest. This question is open in general. However, we have the following result for patterns to allow a nonzero nest.

THEOREM 2.10. *Let A be an $n \times n$ pattern that satisfies the following four conditions*:

(1) *A has at least one nonzero diagonal entry*;
(2) *A allows a nonzero determinant*;
(3) *A is combinatorially symmetric*; *and*
(4) *A is irreducible*.

Then A allows a nonzero nest.

The assumption of combinatorial symmetry cannot in general be relaxed, as the following example shows. The pattern

$$A = \begin{bmatrix} * & * & 0 \\ 0 & 0 & * \\ * & * & 0 \end{bmatrix}$$

satisfies conditions (1), (2), and (4) of Theorem 2.10, but has no nonzero nest.

We now come to our main result.

THEOREM 2.11. *Let the pattern A satisfy the conditions of Theorem* 2.10. *Then the entries of A may be signed so that the resulting sign pattern allows a properly signed nest*.

2.7 CONCLUSIONS

The conditions for a successful qualitative comparative statics analysis have been available for over a decade. The issues remaining concern algorithmic strategies for determining when systems satisfy the conditions, or, perhaps of even more importance, a determination (in some minimal sense) of what additional information beyond sign patterns is required to complete the enterprise of signing the entries of the inverse Jacobian matrix. The issues at stake in utilizing information in addition to sign pattern are considered in the Chapter 3.

Samuelson's query concerning the identification of stable matrices from their sign patterns alone is answered by the Quirk-Ruppert theorem for matrices with negative diagonal entries, and by the Jeffries, Klee, and van den Driessche (1977) theorem for matrices in general, given that the only information available concerning the entries in a matrix A is purely qualitative (sign pattern) information.

For the special case of a 2×2 matrix, the sign stability conditions' holding everywhere (together with the condition $a_{11}a_{22} \neq 0$ or $a_{12}a_{21} \neq 0$

everywhere) implies global stability, from a theorem of Olech (1963). For applications to economics, see Quirk and Ruppert (1967) and Garcia (1972).

There have been a large number of papers in the mathematics literature dealing with the properties of sign stable matrices, with algorithms for identifying such matrices, and with generalizations of the concept of sign stability. Maybee (1974) has used graph-theoretic concepts together with the theory of combinatorially symmetric matrices to develop an alternative proof of the Quirk-Ruppert theorem. Klee and van den Driessche (1977) have stated an efficient algorithm for identifying sign stable matrices. Logofet and Ulyanov (1982a) have given an alternative proof of the Jeffries conditions. Jeffries (1986) has extended the notion of sign stability to nonlinear (nth degree polynomial) systems, and identified sufficient conditions for sign stability of such systems. Yamada (1987) has shown that the original Quirk-Ruppert general-case conditions (i)–(iv) (set forth in Section 2.5) are necessary and sufficient for generic sign stability, i.e., the conditions hold almost everywhere.

Jeffries and van den Driessche (1991) develop a variation of sign stability, here interpreted as stability over certain equivalence classes, to apply to the analysis of linear first-order difference equation systems, and develop necessary and sufficient conditions for their concept of m-stability, again using the notions of cycles and paths in matrices. Eschenback and Johnson (1991) develop necessary and sufficient conditions for sign patterns of matrices such that $B \in Q_A$ has all real characteristic roots, $B \in Q_A$ has all purely imaginary roots, $B \in Q_A$ has all real parts of roots nonpositive, and $B \in Q_A$ has all nonreal roots. Jeffries, Klee, and van den Driessche (1987) and Klee (1989) examine the links among semistability (nonpositive real parts of characteristic roots), quasi-stability (semistable with all roots with zero real part being simple roots), and stability (real parts of roots are negative) in terms of some structural characteristics of matrices, and develop computationally efficient ways to identify semistability, quasi-stability, and stability.

Because sign stable matrices have tree graphs, work on the relationship between trees and characteristic roots (e.g., Jeffries and van den Driessche 1988) has applications to sign stability. Similarly, work by Maybee (1988) and Maybee et al. (1989) on formulas for calculating the determinants of special classes of matrices based on their structural characteristics has potential applications for sign stability research. Finally, Maybee and Voogd (1984) examine the sign solvability aspects of sign stable systems, among other aspects of sign stability.

APPENDIX: SPECIAL TOPICS IN MATRIX ANALYSIS

A2.1 Introduction

As will be seen throughout this book, very few problems are as easily solved as the example in Section 1.4. The conditions for full or partial sign solvability are restrictive. In this appendix, we introduce some facts about matrices, which we use repeatedly in determining sign solvability. These results are not usually found in a first course in linear algebra. Our emphasis is on the notions of irreducible matrices, special formulas for computing determinants, and the concept of a stable matrix.

A2.2 Reducible Matrices

Let $N = \{1, 2, \ldots, n\}$, and suppose A is an $n \times n$ matrix. If $\alpha \subseteq N$ and $\beta \subseteq N$, we will denote by $A[\alpha \mid \beta]$ the *submatrix* of A in rows α and columns β. To each submatrix of A we can assign a real number, called the *size* of $A[\alpha \mid \beta]$ and having the value $|\alpha| + |\beta|$, where $|\alpha|$, for example, means the number of elements in the subset α. When $|\alpha| = |\beta|$, then $A[\alpha \mid \beta]$ is a *square submatrix* of A. To each square submatrix of A we can assign another real number, called the *determinant* of $A[\alpha \mid \beta]$ and written $\det A[\alpha \mid \beta]$. These numbers are called the *minors* of A. For the case in which $\alpha = \beta$, we write $A[\alpha]$ in place of $A[\alpha \mid \beta]$. The submatrix $A[\alpha]$ is called a *principal submatrix* of A, and the number $\det A[\alpha]$ is called a *principal minor* of A. The principal minors of A include $\det A[N]$, written $\det A$, and the case $\alpha = \emptyset$. To $\det A[\alpha]$ we assign the value one when $\alpha = \emptyset$.

The determinant of the $n \times n$ matrix A, $\det A$, consists of $n!$ terms, many of which may be zero if A has some zero entries. If the zero entries are assigned in such a way that each of the $n!$ terms in the expansion of $\det A$ contains a zero entry of A, we will call the matrix A *combinatorially singular*. Equivalent conditions for an $n \times n$ matrix to be combinatorially singular are classical and are contained in the following result.

THEOREM A2.1 (Frobenius 1908). *The $n \times n$ matrix A is combinatorially singular if and only if A has a zero submatrix $A[\alpha \mid \beta] = 0$ of size $|\alpha| + |\beta| \geq n + 1$.*

Notice that this result is consistent with the case most likely to be known to the reader, namely, the case where A has a row or column of zeros. In the case of a zero row, we have $|\alpha| = 1$ and $|\beta| = n$, and in the case of a zero column, we have $|\alpha| = n$ and $|\beta| = 1$.

We will call A a *normalized matrix* if $a_{ii} \neq 0$ for $i = 1, 2, \ldots, n$.

THEOREM A2.2. *The matrix A can be normalized if and only if it is not combinatorially singular.*

PROOF. Obviously if A is a normalized matrix, the product $\prod_{i=l}^{n} a_{ii}$ is a nonzero term in the expansion of det A. Hence A is not combinatorially singular. Conversely, if det A has nonzero terms we can permute the columns of A such that the resulting matrix has one such term on its principal diagonal.

Recall that a matrix P is called a permutation matrix if it is obtained from the identity matrix I by permuting the rows of I. For all such matrices we have det $P = \pm 1$; hence permutation matrices are nonsingular. It is also easy to verify that $P^{-1} = P^T$, i.e., that the inverse of a permutation matrix is its transpose.

Theorem A2.2 implies that a matrix A is not combinatorially singular if and only if there exists a permutation matrix P such that the matrix $B = AP$ satisfies $b_{ii} \neq 0$ for $i = 1, 2, \ldots, n$. Notice that if A is normalized and if $A[\alpha \mid \beta] = 0$, then we must have $\alpha \cap \beta = \varnothing$. Thus, for normalized matrices we can define for $k \geq 0$ the concept of A being k-connected. We say that the $n \times n$ normalized matrix is *k-connected* if the *largest* size-zero submatrix $A[\alpha \mid \beta] = 0$ of A has size $|\alpha| + |\beta| = n - k$. The case where $k = 0$ is classical and goes back to Frobenius's 1912 paper on nonnegative matrices. If A is zero-connected it is called *reducible*. In this case there is a zero submatrix $A[\alpha \mid \beta] = 0$ of size $|\alpha| + |\beta| = n$. Since $\alpha \cap \beta = \varnothing$, we must also have $\alpha \cup \beta = N$. Now we find a permutation matrix P that permutes the indices in α into the set $\{1, 2, \ldots, |\alpha|\}$ and the indices in β into the set $\{|\alpha| + 1, \ldots, n\}$. Then the matrix can be written as

$$B = PAP^T = \begin{vmatrix} B_{11} & 0 \\ B_{21} & B_{22} \end{vmatrix}, \tag{A2.1}$$

where B_{11} is $|\alpha| \times |\alpha|$ and B_{22} is $|\beta| \times |\beta|$. Thus we obtain our next combinatorial result.

THEOREM A2.3 (Frobenius 1908). *The $n \times n$ normalized matrix A is reducible if and only if there exists a permutation matrix P such that the matrix $B = PAP^T$ has the form (A2.1).*

Notice that if A is an irreducible normalized matrix (no such P exists) it must be k-connected for some $k \geq 1$. Hence the largest size-zero submatrix of A is at most of size $n-1$.

For the $n \times n$ matrix A, we call the product $A_p = a_{i(1)i(2)}a_{i(2)i(3)} \cdots a_{i(r-1)i(r)}$ a *path* in A if $r \geq 2$ and the set $V_p = \{i(1), i(2), \ldots, i(r-1)\}$ consists of distinct elements of N. The number $r-1$ is called the *length* of A_p, and the set V_p is called the *index* set of A_p. If we wish to emphasize that $i(1)=j$ and $i(r)=k$, we will write $A_p[j \to k]$. Then j is the *initial index* of the path A_p, and k is the terminal index of the path.

The product $A_c = a_{i(1)i(2)}a_{i(2)i(3)} \cdots a_{i(r)i(1)}$ is called a *cycle* of A if $r \geq 2$ and the product $a_{i(1)i(2)} \cdots a_{i(r-1)i(r)}$ is a path in A. The *length* of the cycle is the number r, and the set $V_c = \{i(1), i(2), \ldots, i(r)\}$ is called the *index* set of the cycle. The diagonal entries $a_{i(q)i(q)}$ can also be considered cycles of A and they have length one and index set consisting of the single index $V_c = \{i(q)\}$. If we wish to emphasize that a cycle A_c contains a particular index j, we write $A_c[j]$.

We use these products throughout the book. One application is to yield another characterization of reducibility.

THEOREM A2.4. *The matrix A is reducible if and only if the set N can be partitioned into two nonempty subsets I and J such that every path $A_p[i \to j] = 0$ for $i \in I$ and $j \in J$.*

This characterization is important because it is a simple matter to determine whether or not a partition exists. On the other hand, since there are $n!$ permutation matrices of order n it is not reasonable to test for reducibility by using partitions.

A2.3 Various Determinant Formulas

Many of the results presented later rely upon formulas for the expansion of determinants of the matrix A using only values of principal minors of A. We remind the reader that there are $n!$ distinct terms in the expansion of det A. Each such term can be written as the product of disjoint cycles of A. (Two cycles $A_{c(1)}$ and $A_{c(2)}$ are disjoint if $V_{c(1)} \cap V_{c(2)} = \emptyset$.) This product is unique up to the order of multiplication. Each cycle A_c contributes a sign, namely, $(-1)^{r+1}$, where r is the length of c, to the term. Thus cycles of even length contribute a negative sign and cycles of odd length contribute a positive sign to the term. It follows that a term t in the expansion of det A can be written as a product,

$$t = A_{c(1)}A_{c(2)} \cdots A_{c(q)}(-1)^s,$$

where s is the number of cycles of even length in the product.

If A_c is a cycle of the $n \times n$ matrix A with index set V_c, then we associate with A_c the principal minor of A, $\det A[N \setminus V_c]$. Here $V_c' = N \setminus V_c$ is the set of indices not belonging to V_c. This set is uniquely determined, and we call the number $\det A[V_c']$ (also uniquely determined) the *cominor* of A_c.

We can now state the Maybee determinantal formula, which is an expansion of $\det A$ using these concepts. To this end, we fix our attention on a particular index $i \in N$. Let $\{A_{c1}[i], \ldots, A_{cm}[i]\}$ be the set of cycles of A that contain the index i in their index sets. Then,

$$\det A = \sum_{j=1}^{m} (-1)^{r(j)+1} A_{ej}[i] \det A[V_{cj}'], \tag{A2.2}$$

where $r(j)$ is the length of the cycle $A_{cj}[i]$. Of course, such a formula exists for each $i \in N$. If A has several entries with the value zero, then some of these expansions may be significantly simpler than others.

The Maybee determinant formula immediately leads to an alternate way of calculating the elements of the inverse matrix. The algebraic cofactor of a_{ij} is the determinant of the submatrix $A[N \setminus \{i\} \mid N \setminus \{j\}]$ and is denoted by A_{ij}. For any path p, let $V_p[i \to j]$ be the index set of this path, and set $V_p(i,j) = N \setminus V_p[i \to j]$ so that $V_p(i,j)$ is the set of indices in N not belonging to the path $A_p[i \to j]$. Set $\delta = N \setminus \{i,j\}$. Let the set of paths in A from index j to index i be $\{A_{p1}[j \to i], \ldots, A_{pu}[j \to i]\}$. Then an application of the formula (A2.2) to the algebraic cofactor of a_{ij} yields the following formula. Setting $A_{ij} = (-1)^{i+j} \det A[\delta \cup \{j\} / \delta \cup \{i\}]$, we have, for $i \neq j$,

$$A_{ij} = \sum_{k=1}^{u} (-1)^{r(k)} A_{pk}[j \to i] \det A\left[V_{pk}(i,j)\right]. \tag{A2.3}$$

Note that V_{pk} contains the indices i and j for all k; hence $A[V_{pk}(i,j)]$ is a principal submatrix of $A[\delta]$. When $i = j$, $\det A[N \setminus \{i\} \mid N \setminus \{j\}] = \det A[N \setminus \{i\}]$, and $(-1)^{i+j} = (-1)^{2i} = 1$. Hence the algebraic cofactor of a_{ii} is

$$A_{ii} = \det A[N \setminus \{i\}]. \tag{A2.4}$$

Now suppose that the matrix A is nonsingular so that $\det A \neq 0$. Then we set $B = A^{-1} = [b_{ij}]$, and it follows from (A2.3) and (A2.4) and the well-known formula for B that for $i \neq j$, we have

$$b_{ij} = \sum_{k=1}^{u} (-1)^{r(k)} A_{pk}[i \to j] \det A\left[V_{pk}(i,j)\right] / \det A, \quad \text{(A2.5)}$$

and

$$b_{ii} = \det A[N \setminus \{i\}] / \det A. \quad \text{(A2.6)}$$

Derivations of the formulae (A2.2), (A2.3), (A2.4), (A2.5), and (A2.6) may also be found in Maybee et al. (1989), but they make extensive use of graph theory.

Alternate versions of the Maybee formulas have been employed in the literature and are restated here for completeness.

Let $\alpha(r, N)$ denote the set of all strictly increasing multi-indices of length r in the index set $N = \{1, \ldots, m\}$, so that the elements of $\alpha(r, S)$ have the form $H = (h_1, \ldots, h_r)$, $1 \leq h_1 < h_2 < \cdots < h_r \leq n$. $A(H)$, $H \in \alpha(r, S)$ denotes the principal submatrix of A with the row and column indices indicated by the indices in H, and A_H is the determinant of this submatrix. Finally, $A_{(H)}$ denotes the sum of all independent cycles of length r in $A(H)$. H' denotes the complement of H in S. Then the Maybee determinant formula is the following for fixed H in $\alpha(n-1, S)$:

$$\det A = a_{H'H'} A_H + \sum_{r=0}^{n-1} (-1)^{n+1-r} \sum_{K \in \alpha(r, H)} A_K A_{(K')}. \quad \text{(A2.7)}$$

An alternative formulation is given by

$$\det A = \sum_{r=0}^{n=1} (-1)^{n+1-r} \left(\frac{n-r}{n} \right) \sum_{H \in \alpha(r, S)} A_H A_{(H')}. \quad \text{(A2.8)}$$

An important byproduct of the Maybee determinant formula is found in determining the signs of terms in the expansion of $\det A$, and under certain circumstances, the sgn of $\det A$ itself. To do this, the matrix is assumed to be normalized (i.e., combinatorially nonsingular) and each column multiplied by -1 as necessary such that $a_{ii} < 0$ for all i. Given this, the term in the expansion of $\det A$ formed from the product of the main diagonal entries will have the sign $(-1)^n$ for A an $n \times n$ matrix.

Consider any other nonzero term, t, in the expansion of $\det A$, expressed as above as the product of elements of A corresponding to disjoint cycles,

$$t = A_{c(1)} A_{c(2)} \cdots A_{c(q)} (-1)^s,$$

where s is the number of cycles with an even number of vertices. When main diagonal entries appear in such terms, each may be considered a cycle of length one. Consider the kth such term, $A_{c(k)}$, where the sign of the cycle itself is negative. If the length of $A_{c(k)}$ is odd, then the sign of the cycle is unmodified, i.e., contributes a positive sign to t, and the term is as if all of its constituent entries were negative. If the length of $A_{c(k)}$ is even, then the term is multiplied by -1, i.e., contributes a negative sign to t. As before, if the sign of $A_{c(k)}$ is negative, then this, too, is as if all of its constituent entries were negative. Accordingly, if all of the cycles that comprise t have negative signs, then $\operatorname{sgn} t = (-1)^n$.

This leads to the following interesting result for qualitative analyses. Let A be an $n \times n$ irreducible matrix with $a_{ii} < 0$ for all i, let $A_{c(k)}$ be the kth cycle of A, and let t be a nonzero element in the expansion of $\det A$; then, $\operatorname{sgn} t = (-1)^n$ for all t if and only if $\operatorname{sgn} A_{c(k)} < 0$ for all k. Clearly, if $\operatorname{sgn} t = (-1)^n$ for all nonzero t, then $\operatorname{sgn} \det A = (-1)^n$. This basic result was first given in Bassett, Maybee, and Quirk (1968), and is fundamental to the algebraic, nonparametric analysis of invertibility in Chapter 3.

A2.4 An Outline of Matrix Stability

The problem of stability of motion occupies a vast portion of the literature of applied mathematics. We cannot hope to cover it here. Accordingly, we discuss the few fundamental results from the theory of matrix stability required in the succeeding chapters.

A matrix is called *stable* if all of its eigenvalues have negative real parts, i.e., if these eigenvalues lie in the open left half of the complex plane. Much of the research on stable matrices is concerned with discovering equivalent conditions for a matrix to be stable. Perhaps the most widely used of these conditions is due to Lyapunov. To state this result, let us recall that the matrix A is *symmetric* if $a_{ij} = a_{ji}$ for all $i \neq j$. A symmetric matrix is called *positive-definite* if all of its principal minors are positive. But when a matrix is symmetric, it is not necessary to verify directly that every principal minor is positive in order to show that it is positive-definite. In fact, if A is symmetric and has a sequence $\{A[\alpha_1], A[\alpha_2], \ldots, A[\alpha_n]\}$ of principal submatrices with $|\alpha_i| = i$ for $i = 1, 2, \ldots, n$, and $\alpha_i \subset \alpha_{i+1}$ for $i = 1, 2, \ldots, n-1$, such that $\det A[\alpha_i] > 0$

for $i = 1, 2, \ldots, n$, then A is positive-definite. Such a sequence is called a *nested sequence* of principal submatrices. We can formalize these results as follows.

THEOREM A2.5. *The symmetric $n \times n$ matrix A is positive-definite if and only if there exists a nested sequence of principal submatrices of A having positive determinants.*

We can now state the Lyapunov theorem.

THEOREM A2.6. *All the eigenvalues of the real matrix A have negative real parts if and only if the matrix equation $A^T B + BA = -I$, where A^T is the transpose of the matrix A, has as its solution B a symmetric positive-definite matrix.*

Observe that Lyapunov's theorem involves solving a system of equations based directly upon the entries of the matrix A. Our second result uses the characteristic polynomial of A, and hence the entries occur only indirectly. Let us write the characteristic polynomial of A in the form

$$p_A(t) = t^n + a_1 t^{n-1} + a_2 t^{n-2} + \cdots + a_{n-1} t + a_n.$$

We can construct the matrix H_n by the following rules: (i) The principal diagonal of H_n is the vector $(a_1, a_2, \ldots, a_n)$. (ii) The kth superdiagonal of H_n is the vector of length $n-k$ with components $(a_{2k+1}, a_{2k+2}, \ldots, a_n, 0, \ldots, 0)$. (iii) The kth subdiagonal of H_n is the vector of length $n-k$ with components $(0, \ldots, 0, 1, a_1, \ldots, a_{n-2k})$. Thus, for example, we have

$$H_5 = \begin{bmatrix} a_1 & a_3 & a_5 & 0 & 0 \\ 1 & a_2 & a_4 & 0 & 0 \\ 0 & a_1 & a_3 & a_5 & 0 \\ 0 & 1 & a_2 & a_4 & 0 \\ 0 & 0 & a_1 & a_3 & a_5 \end{bmatrix}; H_6 = \begin{bmatrix} a_1 & a_3 & a_5 & 0 & 0 & 0 \\ 1 & a_2 & a_4 & a_6 & 0 & 0 \\ 0 & a_1 & a_3 & a_5 & 0 & 0 \\ 0 & 1 & a_2 & a_4 & a_6 & 0 \\ 0 & 0 & a_1 & a_3 & a_5 & 0 \end{bmatrix}.$$

THEOREM A2.7 (Routh 1877; Hurwitz 1895). *All of the eigenvalues of the real matrix A have negative real parts if and only if $\det H_n[1, 2, \ldots, p] > 0$ for $p = 1, 2, \ldots, n$.*

Now we can relate this result to the principal submatrices of A by observing that $a_i = (-1)^i E_i(A)$, $i = 1, 2, \ldots, n$, where $E_i(A)$ is the sum of the principal minors of A of order i.

Observe also that the sequence $\{H_n[1, 2, \ldots, p]\}^p$, of submatrices of H_n is a nested sequence of principal submatrices. On the other hand, it should be observed that H_n is not a symmetric matrix.

We note that $\det H_n[1] = a_1$ and $\det H_n[1,2,\dots,n] = \det H_n = a_n \det H_n[1,2,\dots,n-1]$. Thus we must have $a_1 > 0$ and $a_n > 0$ to satisfy the conditions of Theorem A2.7. This requires, in turn, that we must have $E_1(A) < 0$ and $(-1)^n E_n(A) = (-1)^n \det A > 0$.

A third major result on matrix stability is the theorem of Fisher and Fuller, which also involves the existence of a nested sequence of principal minors.

THEOREM A2.8 (Fuller and Fisher 1958). *Let A be a real matrix having a nested sequence of principal minors $\{A[\alpha_1],\dots, A[\alpha_n]\}$ such that sign $\det A[\alpha_i] = (-1)^i$, $i = 1,2,\dots,n$. Then there exists a diagonal matrix $D = \operatorname{diag}(d_1, d_2, \dots, d_n)$ with $d_k > 0$ for $k = 1,2,\dots,n$ such that DA is a stable matrix.*

Theorems A2.5 through A2.8 furnish important tools for investigating the stability of matrices even though they may often be difficult to apply in specific cases.

There are special types of stability problems that occur in economics and other sciences where it may be difficult or impossible to obtain accurate numerical values for all or even some of the elements in the matrix of a linear system. To understand these problems we now require some qualitative ideas.

By the *sign pattern* of a real matrix A we mean the matrix $\operatorname{sgn} A = [\operatorname{sgn} a_{ij}]$. (Recall that when x is a real number, $\operatorname{sgn} x = 1$ if $x > 0$, $\operatorname{sgn} x = 0$ if $x = 0$, and $\operatorname{sgn} x = -1$ if $x < 0$). Two matrices A and B have the *same sign pattern* if either $a_{ij}b_{ij} > 0$ or both a_{ij} and $b_{ij} = 0$ for all i and j. We say that the matrix A is *qualitatively invertible* (a *QI-matrix*) if every matrix B such that A and B have the same sign pattern is nonsingular. Similarly we call the matrix A *sign stable* if every matrix B such that A and B have the same sign pattern is a stable matrix. The problems of characterizing the sign patterns of QI-matrices and of sign stable matrices have been solved and form much of the subject matter of Chapters 2 and 3.

There is a problem closely related to the problem of sign stability, which is also important in the applications. We call the matrix A a *potentially stable matrix* if there exists a stable matrix B such that A and B have the same sign pattern. The problem of characterizing the sign patterns of potentially stable matrices is unsolved at present.

It is easy to give an example of a matrix that is both a QI-matrix and a sign stable matrix. Let a, b, and c be positive numbers, and let

$$A = \begin{bmatrix} -a & b \\ -c & 0 \end{bmatrix}.$$

Then we have

$$A^{-1} = \frac{1}{bc}\begin{bmatrix} 0 & -b \\ c & -a \end{bmatrix},$$

and $\det(\lambda I - A) = \lambda^2 + a\lambda + bc$, which has zeros $\lambda_1 = 1/2(-a + \sqrt{a^2 - 4bc})$, both of which have negative real parts for all positive values of a, b, and c. Hence the sign pattern,

$$\begin{bmatrix} - & + \\ - & 0 \end{bmatrix}$$

is that of both an QI-matrix and a sign stable matrix.

3

Information and Invertibility

3.1 INTRODUCTION

The basic issue addressed in this chapter may be summarized as follows: *For the linear system $Ax = y$, how do different states of knowledge about the entries of the matrix A enable conclusions to be reached about characteristics of the matrix A^{-1} and the stability of A?*

An important matter to resolve in approaching this problem is a determination of what is to be meant by "knowledge about the entries of the matrix A." Starting with Chapter 5, the applied framework for which the model was developed, i.e., equilibrium or optimization, will be an important source of information about the entries of the Jacobian matrix. In this chapter, the application to which the model is being applied will not be identified. Instead, the problem will be put on a purely informational basis concerning A's entries: *Given a specific set of facts about the entries of A, what can be said about the entries of A^{-1} and the stability of A?*

In approaching this problem the analysis methodology will be based upon two rather different analytical techniques. For all of the cases considered, an *algebraic* method will be used that enables the determination of correspondences between terms in the expansion of a matrix's determinant and characteristics of the (signed) directed graph corresponding to the matrix. This approach was utilized in Chapter 2, and its development and results will be reiterated here in Section 3.2 as part of the development of methods to utilize different types of information about the entries of A. Advantages of the algebraic approach can be found in the insights gained about the nature of the matrix being analyzed and (at least sometimes) efficiencies in detecting when certain conditions on a matrix are satisfied, even for relatively large arrays. Difficulties in applying the approach can be found in the need to transform the matrix and perhaps its submatrices into a standard form and in the fact that developing and applying algorithms that enable the approach to be implemented presents a nontrivial technical burden.

An alternative approach is the *elimination* method described in Section 3.3. This approach amounts to implementing an algorithmic strategy to detect characteristics of A^{-1} through inspection. No transformations of the matrix's form or preconditions on its structure are required, and the algorithmic principles involved are straightforward. The difficulty in using the elimination method relates to its efficiency. For a standard qualitative problem (e.g. finding the conditions under which the specification of sgn A assures that det $A \neq 0$), if used without restrictions on A's structure, the elimination method requires that 3^n-many solution vector sign patterns be tested against the chance that only one of them is valid. For the case of a 20×20 linear system, this would involve testing almost 3.5 billion sign patterns. Further, the method would have to be applied to n-many similar problems in order to resolve the signs of all of the entries of the inverse matrix (around 70 billion sign patterns in all). Still, the elimination method is particularly handy for sufficiently small problems, and illustrations of its application are provided.

To make progress beyond the results obtained in the Sections 3.2 and 3.3 requires some kind of categorization of "knowledge about the entries of the matrix A." The analytical approach will be to deal first with the invertibility of A and determinable characteristics of the entries of A^{-1}. Section 3.4 introduces the idea of the hierarchical structure of measurement scales (i.e., nominal, ordinal, interval, ratio). The general idea is to specify ever more restrictive information categories, with each new category providing the information of the immediately prior category plus more information, and then developing conditions under which properties of the inverse matrix can be found on the basis of the information assumed to be available. To the degree to which characteristics of the inverse matrix can be based upon less, rather than more, information about the matrix A, the underlying model is more robust and probably can be more readily tested (i.e., refuted).

In Section 3.5, the invertibility of a matrix based only upon information about the relative size (i.e., ranking) of its entries will be studied. In Section 3.6, information in addition to a ranking of A's entries will be considered. A summary of the information hierarchy developed and the associated conditions for invertibility will be given in Section 3.7. The stability of A will be related to the information categories in Section 3.8.

3.2 AN ALGEBRAIC ANALYSIS OF QUALITATIVE INVERTIBILITY

The specific problem to be studied in this section and the sections through 3.7 is the question, *Under what circumstances can given information about the entries of A be shown to ensure that* det$A \neq 0$?

Usually, but not always, A will be transformed such that

$$\text{if } \det A \neq 0, \text{ then sgn} \det A = (-1)^n.$$

As the analysis is developed for different cases, a hierarchical categorization of information about the entries of the matrix A will be developed in parallel.[1]

Matrix Structure and the Directed Graph Corresponding to a Matrix

A square matrix can be interpreted to represent a pattern of influence among the variables that correspond to each of its columns. One manner of doing this is to construct and inspect the (signed) directed graph that corresponds to the matrix. Let A be the $n \times n$ coefficient matrix for the linear system $Ax = y$, and let

$$\text{DG}(A) = \{(i): i = 1, 2, \ldots, n; \text{``} \rightarrow \text{''s}\}$$

be the directed graph corresponding to A, where the n-many is are vertices, each distinctly corresponding to an entry of the solution vector x; and, the arrows are directed arcs among pairs of vertices such that $j \rightarrow i$ if and only if $a_{ij} \neq 0$. For the directed arc $j \rightarrow i$, j is called the initial vertex and i is called the terminal vertex of the arc.

To make an interpretation about "patterns of influence" among the variables of the system, the assignment of indices to the entries of A must be such that A is put into a standard form (the first of several standard forms). Define A as being in Standard Form 1 (SF1) if and only if the indices of A have been assigned such that $a_{ii} \neq 0$ for all i. From Theorem A.2.2, this means that the matrix is not combinatorially singular. The ability to bring a matrix into this form can be related readily to the algorithm for computing its determinant. Consider that the determinant of a matrix is the sum of $n!$-many distinct terms, with each term comprised of the product of n-many entries of the matrix such that each row and each column of the matrix are represented exactly once. A matrix is invertible only if at least one of these terms is nonzero. Consider any such nonzero term with its constituent entries arranged in the natural order of rows (i.e., $a_{1*}a_{2*} \cdots a_{n*}$). To the degree necessary, reindex the columns of the matrix so that the column index of each entry in the term equals the row index. The result of the index transformation is to bring the nonzero term onto the maindiagonal of the transformed matrix, and hence puts the matrix into SF1. As a practical matter, the renumbering of columns simply amounts to reindexing variables; and given this, it does not add to or detract from the mathematical properties of the system.

For a matrix in SF1, each variable appears in the equation with the same index (among the equations in which it appears). This equation may be interpreted as the equation corresponding to the variable. In fact, many systems can be expressed in a fashion that immediately presents a correspondence between variables and equations, i.e., each equation can be written as

$$x_i = f^i(x_j : j \neq i; y),\ i = 1, 2, \dots, n.$$

In this regard, for equation $\#i$, if variable $\#j$ appears in the equation (i.e., if a_{ij} [can be] $\neq 0$), then one could make the interpretive statement, Variable $\#i$ "depends on" variable $\#j$. Operationally, this circumstance refers to the fact that the system cannot be solved for the value of variable $\#i$ unless it is also solved for the value of variable $\#j$. "Dependence" in this sense is a transitive, binary relation among variables. Given transitivity, the dependence relation (variable $\#i$ depends on variable $\#j$) may also apply for $a_{ij} = 0$, if there is a sequence of distinct directed arcs (i.e., a sequence of nonzero coefficients), $(j) \to (g) \to \cdots \to (h) \to (i)$. Such a sequence is termed a *path* in DG(A) and notated $p(j \to i)$.

For any matrix A, its *Boolean image*, $B = [b_{ij}] = \text{bol}\, A$, could be formed such that $b_{ij} = 1$ if and only if $a_{ij} \neq 0$ and $b_{ij} = 0$ otherwise. For the sake of parallelism to more restrictive cases developed below, let B_A stand for the class of matrices with the same Boolean image as A. From the discussion above, a specification of bol A is made by a determination of which variables appear in which equations. Even this basic specification of the mathematical model being studied (e.g., about the interrelationships in the real system being modeled) can lead to useful conclusions about A^{-1}. Specifically, it is possible for det $A \neq 0$ only if the expansion of det A has at least one nonzero term: det $A \neq 0$ only if bol A can be put into SF1. Call such matrices BI-matrices (BI for *Boolean invertible*). Further, for bol A in SF1, the circumstance $a_{ij}^{-1} \neq 0$ expresses a "transmission of influence" from variable $\#j$ (i.e., from a disturbance of the parameters in equation $\#j$, which corresponds to variable $\#j$) to variable $\#i$ (i.e., the disturbance potentially leads to a change in the solution value of x_i). From the discussion, this is only possible if a path $p(j \to i)$ exists, a circumstance that can be determined from an inspection of DG(A).[2]

Taken together, the results of this discussion provide for an initial information category, with its implications for invertibility:

Information Category I. A specification of which variables appear in which equations.
Class of matrix identified. The incidence matrix bol $A \in B_A$.
Determinable characteristics of A^{-1}
Category I.1 (*invertibility*). det $A \neq 0$ only if A can be put into SF1.
Category I.2 (*nonzero entries*). For A in SF1, $a_{ij}^{-1} \neq 0$ only if a $p(j \to i)$ exists in DG(A).

For A in SF1, the facts available from information category I can be processed to reveal the "structure" of the linear system. Let *structure* refer to a specification of which variables of the linear system must be solved for simultaneously versus which variables can be solved for independent of one another. It should be clear that if a $p(j \to i)$ exits in DG(A), then the value of variable $\#j$ must be found in order for the value of variable $\#i$ to be found. If a $p(i \to j)$ also exists, then the values of variables $\#i$ and $\#j$ must be solved for simultaneously. Let

$$V = \{(i) \colon i = 1, 2, \ldots, n\}$$

be the vertex set of DG(A) and $V_1 \subseteq V$. Then V_1 is a *component* of DG(A) if and only if for any $\{i, j\} \in V_1$, $p(i \to j)$ and $p(j \to i)$ exist in DG(A). Given this, V_1 is called a *strong component* of DG(A) if and only if for any other V_j a component, either $V_J \subseteq V_I$ or $V_I \cap V_J = \{\phi\}$. The point of the discussion is that the components of DG(A) identify the variables of the linear system that must be solved for simultaneously. It is clear from the discussion that if DG(A) has m-many strong components, then the strong components taken together are a partition of V. Further, for

$$V = \{V_1, V_2, \ldots, V_m),$$

the strong components of DG(A), it is possible to assign indices to the strong components such that for $i \in V_g$, $j \in V_h$, and $g < h$, then the path $p(j \to i)$ does not exist in DG(A). Operationally, the organization of structure in this way reveals that the variables identified by the strong components can be solved for, group by group, with the values found for variables in strong components with lower indices independent of the values found for variables in strong components with higher indices, but not (necessarily) conversely. Call an assignment of indices to the strong components of DG(A) *structured* if the indices identify a group by group solution sequence as described above. Given this, for components with structured indices, if $i \in V_g$, $j \in V_h$, and $g < h$, then $a_{ij}^{-1} = 0$ for all $A \in B_A$.

The significance of structure in this sense is that, on the one hand, it enables a determination of entries of A^{-1} that must be zero (*logical zeros*), while, on the other hand, it imposes a severe burden on the derivation of results by means of the algebraic techniques being developed. Specifically, the results found must be first applied to the submatrices of A corresponding to the strong components of DG(A). Once done, the structure of the components themselves (i.e., the many ways that structured indices might be assigned to the strong components) must be accounted for in order to extend the results to the matrix A itself. Systems found to be "partially invertible" are often such that invertibility conditions are satisfied for some of the submatrices corresponding to strong components, but not for all such submatrices.

In this chapter, problems of this kind will be avoided by assumption. If DG(A) has more than one strong component, call A *reducible*. If all of the vertices, V, are themselves a component in DG(A), then call A *irreducible*. For A irreducible, all of the variables of the linear system must be solved for simultaneously. In general, the matrices considered here will be assumed to be irreducible.[3] Given this, no entries of the inverse matrix will be necessarily zero. The reader can confirm that the matrix below is reducible, with strong components with structured indices given by $V_1 = \{1,4\}$ and $V_2 = \{2,3\}$:

$$\text{bol } A = \begin{bmatrix} 1 & 0 & 0 & 1 \\ 1 & 1 & 1 & 0 \\ 0 & 1 & 1 & 0 \\ 1 & 0 & 0 & 1 \end{bmatrix}.$$

Alternatively, the matrix below is irreducible:

$$\text{bol } A = \begin{bmatrix} 1 & 0 & 0 & 1 \\ 1 & 1 & 0 & 0 \\ 0 & 1 & 1 & 0 \\ 0 & 0 & 1 & 1 \end{bmatrix}.$$

Qualitative Invertibility and the Signed Directed Graph Corresponding to a Matrix

The directed graph corresponding to a matrix can be easily "signed" by adding information about the signs (positive or negative) of A's nonzero entries. Let

$$\text{SDG}(A) = \{(i)\colon i = 1,2,\ldots,n; \text{“} \rightarrow_{+} \text{”'s and “} \rightarrow_{-} \text{”'s}\}$$

be the signed directed graph corresponding to A, with $(j) \rightarrow_{+} (i)$ if and only if $a_{ij} > 0$ and $(j) \rightarrow_{-} (i)$ if and only if $a_{ij} < 0$. Paths in SGD(A) can also be signed. Let $\operatorname{sgn} p(j \rightarrow i)$ be the sign of the product of the coefficients corresponding to the arcs in $p(j \rightarrow i)$. For a given path $p(j \rightarrow i)$, if also $a_{ji} \neq 0$, then there is a path from vertex (i) to itself. Call such a circumstance a *cycle*, $c(i)$. As with paths, the sign of a cycle, $\operatorname{sgn} c(i)$, is the sign of the product of the coefficients corresponding to the arcs in $c(i)$.

Following the above discussion, for any matrix, its "sign pattern," $B = [b_{ij}] = \operatorname{sgn} A$, could be formed such that

$$b_{ij} = 1 \text{ (resp.} -1) \text{ iff } a_{ij} > 0 \text{ (resp. } a_{ij} < 0),$$

and

$$b_{ij} = 0 \text{ iff } a_{ij} = 0.$$

Given this, let Q_A stand for the class of matrices with the same sign pattern as A. The information category represented by sgn A is clearly hierarchical compared to bol A since

$$\text{bol}\{\text{sgn } A\} = \text{bol } A.$$

Conditions for the invertibility of a matrix based only upon sgn A are well established and given in Theorem 2.1. For the algebraic method being developed here, the conditions require that the form of A be further transformed. Specifically, for A in SF1, A is put into Standard Form 2 (SF2) by multiplying the columns of A by -1 as necessary such that $a_{ii} < 0$ for all i. This transformation does not add to (or detract from) the invertibility of A; instead, if A can be otherwise shown to be invertible, then the transformation changes the sign of det A if the number of columns multiplied by -1 is odd in number. Given a matrix sign pattern, the conditions for invertibility may be summarized as follows:

Information Category II. The same as category I plus a specification of the sign (positive or negative) of the nonzero entries of A.

Class of matrix identified. The matrix of signs $\operatorname{sgn} A \in Q_A$.

Determinable characteristics of A^{-1}[4]

Category II.1 (*invertibility*). For A an irreducible $n \times n$ matrix in SF2, $\det B \neq 0$ for all $B \in Q_A$ if and only if $\operatorname{sgn} c(i) < 0$ for all cycles in SDG(A). Call such matrices QI-matrices (QI for qualitatively invertible). If A is an irreducible $n \times n$ *QI*-matrix in SF1, then the sign of the determinant of A is given by sgn det $A =$

$(-1)^z(-1)^n$, where z is the number of columns multiplied by -1 to put A into SF2.

Category II.2 (necessarily determinable entries). For A an irreducible $n \times n$ matrix in SF1, if A is a QI-matrix when put into SF2, then $\operatorname{sgn} a_{ij} \neq 0$ implies $\operatorname{sgn} a_{ji}^{-1} = \operatorname{sgn} a_{ij}$.

Category II.3 (possibly determinable entries). For A an irreducible $n \times n$ QI-matrix in SF2, if $\operatorname{sgn} a_{ij} = 0$, then $\operatorname{sgn} a_{ji}^{-1}$ is determinable if and only if all of the paths $p(i \to j)$ have the same sign. If all of the $\operatorname{sgn} p(i \to j)$ have the same sign, then

$$\operatorname{sgn} a_{ji}^{-1} = -\operatorname{sgn} p(i \to j).$$

Category II.4 (entries of sgn A^{-1} that cannot be determined). For A an irreducible $n \times n$ QI-matrix in SF2, if $a_{ij} = 0$, then $\operatorname{sgn} a_{ji}^{-1}$ cannot be determined if and only if A is not a QI-matrix for $a_{ij} \neq 0$. For this case, call $a_{ij} = 0$ a *necessary zero* in sgn A for A a QI-matrix.

The matrix below is a QI-matrix. None of the zero entries are necessary zeros. As a result, all entries of the inverse matrix can be signed.

$$\text{For sgn } A = \begin{bmatrix} -1 & 0 & 1 \\ -1 & -1 & 0 \\ 0 & 1 & -1 \end{bmatrix}, \text{sgn } A^{-1} = \begin{bmatrix} -1 & -1 & -1 \\ 1 & -1 & 1 \\ 1 & -1 & -1 \end{bmatrix}.$$

The matrix below is also a QI-matrix; however, the single zero entry is a necessary zero (the reader can confirm this). As a result, the transposed entry of the inverse cannot be signed.

$$\text{For sgn } A = \begin{bmatrix} -1 & 1 & 1 \\ -1 & -1 & -1 \\ 0 & 1 & -1 \end{bmatrix}, \text{sgn } A^{-1} = \begin{bmatrix} -1 & -1 & * \\ 1 & -1 & 1 \\ 1 & -1 & -1 \end{bmatrix}.$$

3.3 THE ELIMINATION PRINCIPLE

The elimination principle is due to Lancaster (1966). Application of the principle has been refined and extended by Ritshard (1983) and Gillen and Guccione (1990). The analytical principle at stake is extremely straightforward. For the linear system $Ax = y$, about which only the sign patterns sgn A and sgn y are known, it is possible from inspection to determine if a given sign pattern sgn x could possibly correspond to the sign pattern of the solution to the system. Specifically, the kth row of signs $\{\text{sgn } a_{kj}x_j\colon j = 1, 2, \ldots, n\}$ is "consistent" with a given sign sgn y_k if and only if

(i) if sgn $y_k \neq 0$, then sgn $a_{kg}x_g = \text{sgn } y_k$ for some g; or
(ii) if sgn $y_k = 0$, then either sgn $a_{kj}x_j = 0$ for all j or there are g and h such that sgn $a_{kg}x_g = -\text{sgn } a_{kh}x_h \neq 0$.

For any proposed solution sign pattern, sgn x, if these conditions are not satisfied for every row of $Ax = y$, then the proposed sign pattern can be *eliminated* as impossible. Characteristics of the solution to $Ax = y$ that can be derived from sign pattern information alone can thus be determined by inspection. If all possible solution sign patterns are inspected and inconsistent solution sign patterns are eliminated, then if the sign of a solution variable is the same for all sign patterns that are not eliminated, the matrix A^{-1} has entries that can be signed on the basis of sgn A. In its most general form, the matrix A need not be transformed for the elimination principle to be applied. Further, the use of the principle is not made more complex when A is reducible versus irreducible; however, A's structure does play a role in the algorithmic implementation of the procedure. Specifically, if A is irreducible, then solution sign patterns with zero entries do not need to be tested (i.e., if a solution with a given entry equal to zero were not eliminated, then solutions with any sign for the given entry, but otherwise identical, would also not be eliminated). Given this, "only" 2^n-many solution sign patterns need to be tested. Alternatively, if A's structure is not known (or is reducible), then there can be a solution variable that is necessarily zero. In this case, 3^n-many solution sign patterns must be tested. As noted above, the number of possible sign patterns to test can quickly become very large. Nevertheless, the elimination principle is a handy and practical approach to investigating the qualitative and (as shown in the next section) other nonparametric properties of a linear system.

As an example, consider the (irreducible) matrix sgn A shown below. For this example, the sgn y column has been set with the first entry positive and the rest zero. Each possible solution sign pattern is enumerated and tested for consistency, row by row. If the sign pattern is inconsistent, the first row for which this is detected is identified under the header "Elimination Row." For the example, application of the elimination principle reveals that only one solution sign pattern of the sixteen possible sign patterns ($= 2^4$, with no zeros) is consistent with all of the rows of the given $Ax = y$. The single consistent sign pattern is identified in the last row. For this example, since only the first entry of sgn y was nonzero and positive, the single consistent sign pattern is the first column of sgn A^{-1} (and A is shown to be a QI-matrix). Other columns of sgn A^{-1} could be investigated similarly by setting the appropriate entry of sgn y positive, and the rest zero.

ROW	sgn A	sgn y
1	−1 −1 −1 −1	1
2	1 −1 −1 −1	0
3	0 1 −1 −1	0
4	0 0 1 −1	0

SOLUTIONS[5]

SOLN#	sgn x Tested	Elimination Row
0	−1 −1 −1 −1	None
1	1 −1 −1 −1	ROW#2
2	−1 1 −1 −1	ROW#3
3	1 1 −1 −1	ROW#3
4	−1 −1 1 −1	ROW#4
5	1 −1 1 −1	ROW#4
6	−1 1 1 −1	ROW#4
7	1 1 1 −1	ROW#4
8	−1 −1 −1 1	ROW#4
9	1 −1 −1 1	ROW#4
10	−1 1 −1 1	ROW#4
11	1 1 −1 1	ROW#4
12	−1 −1 1 1	ROW#3
13	1 −1 1 1	ROW#3
14	−1 1 1 1	ROW#2
15	1 1 1 1	ROW#1
SOLN X	−1 −1 −1 −1	Decidable Variables

For this particular sgn A, all of the entries of the first column of sgn A^{-1} can be derived from sgn A. As a result, since the matrix is in SF1 (actually in SF2), category II.1 can be applied. Given this, the entries in sgn A^{-1} incident on the nonzero entries of the transpose of sgn A are qualitatively decidable, and equal to the transposed entries. Thus, the success of the elimination principle for (any of) the entries of the first column of sgn A^{-1} in combination with category II.2 enables the determination of all but three entries of sgn A^{-1}:

$$\operatorname{sgn} A^{-1} = \begin{bmatrix} -1 & 1 & ? & ? \\ -1 & -1 & 1 & ? \\ -1 & -1 & -1 & 1 \\ -1 & -1 & -1 & -1 \end{bmatrix}.$$

The entries identified by question marks require further analysis. For this matrix, these entries correspond to transposed *necessary zeros*. As a result, from category II.4 these entries of A^{-1} cannot be derived from sgn A. This is illustrated below by using the elimination principle to investigate (say) the last column of sgn A^{-1}.

ROW	sgn A	sgn y
1	−1 −1 −1 −1	0
2	1 −1 −1 −1	0
3	0 1 −1 −1	0
4	0 0 1 −1	1

SOLUTIONS

SOLN#	sgn x Tested	Elimination Row
0	−1 −1 −1 −1	ROW#1
1	1 −1 −1 −1	ROW#2
2	−1 1 −1 −1	ROW#3
3	1 1 −1 −1	ROW#3
4	−1 −1 1 −1	None
5	1 −1 1 −1	None
6	−1 1 1 −1	None
7	1 1 1 −1	None
8	−1 −1 −1 1	ROW#4
9	1 −1 −1 1	ROW#4
10	−1 1 −1 1	ROW#4
11	1 1 −1 1	ROW#4
12	−1 −1 1 1	ROW#3
13	1 −1 1 1	ROW#3
14	−1 1 1 1	ROW#2
15	1 1 1 1	ROW#1
SOLN X	* * 1 −1	Decidable Variables

Note: Asterisk indicates a qualitatively undecidable variable.

For this second example there are four consistent solution sign patterns that were not eliminated. From inspection it may be seen that for all of these, sgn $x_3(=1)$ and sgn $x_4(=-1)$ have the same sign. Accordingly, the corresponding entries of sgn A^{-1} are determinable from sgn A. The remaining entries of sgn A^{-1} can be investigated similarly by setting sgn y_2 (resp. sgn y_3) positive, setting the remaining entries of sgn y at zero, and then repeating the analysis. For the given sgn A (a QI-matrix) the entries of sgn A^{-1} would be found to be as given below:

$$\operatorname{sgn} A^{-1} = \begin{bmatrix} -1 & 1 & * & * \\ -1 & -1 & 1 & * \\ -1 & -1 & -1 & 1 \\ -1 & -1 & -1 & -1 \end{bmatrix}.$$

Although convenient for small systems, the elimination principle becomes increasingly difficult to apply as the dimension of the system being studied increases. Even though A's structure need not be considered, the difference between testing 2^n-many solutions for A known to be irreducible and testing 3^n-many solutions for A's structure reducible or unknown can become very significant (for $n = 20$, $2^n = 1{,}048{,}576$ and $3^n = 3{,}486{,}784{,}401$). In the next section, the analysis of matrices that are not QI-matrices is conducted on the basis of additional information about the relative size of the matrix's entries. For certain kinds of problems of this form, the elimination algorithm can be used effectively to process the additional information.

3.4 CATEGORIZING INFORMATION

A sensible structure for analysis with less than fully quantitative information is to investigate what properties of the inverse Jacobian matrix can be determined on the basis of the type of measurement that is assumed to be made for the entries of the Jacobian matrix. *Measurement* is the assignment of numbers to entities or elements of a class. Different types of measurement are distinguished by the attributes of the numbers assigned that correspond to attributes of the entities, i.e., the attributes that are used by the measure. The three more restrictive types of measurement scale, *ordinal*, *interval*, and *ratio*, appear in economic theory.[6]

Nominal measurement represents an equivalence relation between the entities being considered. The numbers assigned are significant only with respect to the equal and unequal relationships among them. The equivalence relation is complete and transitive. Given an assignment of numbers that comprises a nominal scale, the nominal measurement is preserved if the assignment of numbers is changed as long as entities assigned equal numbers continue to be assigned equal numbers, and conversely. For this reason, nominal measurement is said to be unique up to a one-one onto transformation of scale values (i.e., any such transformation retains the same measurement and is equivalent to the measures being transformed). Social security numbers are an example of nominal measurement.

Ordinal measurement represents a complete, transitive, asymmetric, binary relation among the entities being considered. It is typical to assign numbers such that a weak inequality relation among the numbers assigned, say equal or greater, is homomorphic with the binary relation being measured. This relation is termed a "preordering." Ordinal measurement is often used to establish a rank ordering among the entities being measured. For x an n-vector and $u(x)$ an ordinal measurement of x, $T(u(x))$ is also an ordinal measurement of x provided T is an increasing monotonic transformation of the original scale values. Grades of rank within the General Schedule of civil service employment (e.g., GS-5, GS-6, and so on) are an example of ordinal measurement.

Interval measurement adds a unit of measure to an ordinal scale. In addition to a rank order, interval measurement imposes uniqueness among the ratios of differences between scale values (an ordinal measurement can represent that "a" is ranked before "b," but there is no measure of how much more). The interval measure introduces the notion of "how much." It does so by the values assigned to the ratios of differences between entities. For any interval measure $u(\cdot)$ of entities a, b, c any measure preserving transformation of the scale, $T[u(\ )]$, must satisfy

$$\frac{T[u(a)] - T[u(b)]}{T[u(b) - T[u(c)]} = \frac{u(a) - u(b)}{u(b) - u(c)}.$$

Given this, interval measurement is unique up to an increasing linear transformation, i.e., $T[u(\cdot)] = \alpha u(\cdot) + \beta,\ \alpha > 0$. Measures of temperature in degrees Fahrenheit or degrees Celsius are interval measures. It is a secondary school exercise to express one measure as a linear transformation of the other, e.g., $F^\circ = (9/5)C^\circ + 32$. Note that the typology of scales being developed is cumulative and increasingly restrictive. Interval scales are also ordinal scales, and ordinal scales are also nominal scales.

Ratio measurement adds the assignment of an origin to an interval measure; i.e., the element to which zero is assigned is unique and must be the same for any equivalent ratio measurement. For ratio measurement, the ratio of the scale values themselves is unique (hence the terminology). Ratio measures possess the characteristics of real numbers and are said to be unique up to multiplicative transformations: $T[u(\cdot)] = au(\cdot)$, $a > 0$. Ratio scales are also interval scales.

These information categories are useful, but will be amended somewhat as applied here. For the Boolean and qualitative cases so far developed, researchers can usually specify which entries of the Jacobian matrix are zero and which entries are not zero. The property zero, or nonzero, applied to A's entries establishes two equivalent classes. This information amounts to the specification of which variables appear in which equations. For the Jacobian matrix A, such information is termed here the *Boolean image* of A. If A is reducible, then all the members of the class of matrices with the same Boolean image as A have the same zero entries in their inverse as does A. Accordingly, a specification of the Boolean image of the Jacobian can reveal that the solution values of some of the endogeneous variables are independent of changes in certain of the assumed parameter values. A specification of bol A is a nominal measure in the sense of identifying equivalence classes; however, as in ratio measures, the assignment of zero has a numerical significance.

The next class of information available to researchers is typically a knowledge of A's sign pattern. The *qualitative image* of A, sgn A, involves a unique assignment of zero, as in ratio measurement, with the remaining elements ranked relative to zero, i.e., bigger (positive) or smaller (negative). As such, it does not fit neatly into the hierarchy of scales outlined above. The groupings into positive and negative elements is on the one hand nominal in that these groups constitute equivalence classes, but, on the other hand, there is the ranked relationship of these groups with respect to elements to which zero is assigned. As a result, a specification of sgn A constitutes a hybrid measure compared to the purely cumulative hierarchy of scales.

Subject to the assignment of zero and the groupings with respect to sign, a next class of (additional) information is a ranking of the absolute values of entries in A. In order to sign the entries of the inverse Jacobian matrix, such extraqualitative information may only be required about some, but not all, of the entries of sgn A (as shown below). The "extreme" case is that of a complete ordinal ranking of all of the entries of sgn A. Following the organizing principles above, all matrices with the same ordinal ranking of entries as A can be said to be in the class R_A (R for "ranked"). Sufficient conditions for the invertibility of members of R_A would then depend upon an ordinal measurement of the entries of A, subject to the distinct assignment of zero. Of additional interest are intermediate cases for which some, but not all, ranking relationships of the entries of A are needed to resolve elements of sgn A^{-1}. Some of these intermediate cases are also briefly discussed.

Short of a full quantification of the entries of A (ratio measure), is the concept of interval measure. In the theory of scales, this class of measure involves the specification of a unit of measure without the assignment of a distinct origin. For the information hierarchy at issue here, a distinct origin is already assigned for the specification of the Jacobian matrix sign pattern. Accordingly, the interval concept for the conditions developed here will relate to upper and lower bounds within which the values of entries will be assumed to lie. This is a natural application of error bounds in cases where the Jacobian matrix entries are evaluated through statistical procedures. As such, this final category of information does not correspond readily to the hierarchy of scales.

In summary, the hierarchy of categories that specify the information assumed to be known about the entries of the Jacobian matrix are as follows:

Category 1 (*bol* A, *the Boolean image of* A*—nominal measurement, plus the assignment of zero*). Zero versus nonzero entries are identified.

Category 2 (*sgn* A, *the sign pattern of* A*'s entries—hybrid measurement*). The information from category 1, plus the signs of the nonzero entries.

Category 3 (*sgn* A, *plus some ranking information—hybrid measurement*). The information from category 2, plus some ranking information sufficient to determine some entries in sgn A^{-1}.

Category 4 (*ord* A *as defined below—ordinal measurement of* A*'s entries plus an origin*). The information from category 2, plus ranking information for all entries of A.

Category 5 (*int*{U, V} A, *as defined below—ratio measurement with upper and lower bounds*). Ratio measures of A's entries plus upper and lower bounds (U and V, respectively).

3.5 THE INVERTIBILITY OF MATRICES WITH RANKED ENTRIES

The conditions for the invertibility of sgn A are very restrictive. It is reasonable to expect that many applied systems will not have signed Jacobian matrices that satisfy the conditions.[7] At issue is the need for, and manner of using, additional information about the entries of the Jacobian matrix. The next class of information to be considered will be the relative size of the entries of A. For any matrix A, let $B = [b_{ij}] =$ ord A be such that

$$\text{sgn } B = \text{sgn } A,$$

and

$$b_{ij} > b_{gh} \text{ iff } a_{ij} > a_{gh}.$$

Given A, ord A is a matrix with entries formed by any sign and order preserving transformation of A. For any matrix A, let R_A be the class of matrices with entries formed by sign and order preserving transformations of the entries of A. The specific topic of this section is the conditions on ord A that ensure the invertibility of all of the members of R_A.

The use of the relative size of the entries of A in studying invertibility has been recognized as a natural adjunct to the elimination principle (e.g., Lancaster 1966; Ritschard 1983). In fact, the need for information in addition to sgn A could be selective and not require a full specification of ord A. As an example, consider the 3×3 matrix sgn A given below. This matrix is not a QI-matrix, and (therefore) there are no qualitatively decidable variables for the case tested: sgn $y = (1, 0, 0)$. For this (or any other sign pattern for sgn y), both an all-negative and an all-positive solution vector are qualitatively consistent (i.e., are not eliminated).

ROW	sgn A			sgn y
1	−1	1	0	1
2	1	−1	−1	0
3	0	1	−1	0

SOLUTIONS

SOLN#	sgn x Tested			Elimination Row
0	−1	−1	−1	None
1	1	−1	−1	ROW#1
2	−1	1	−1	ROW#3
3	1	1	−1	ROW#3
4	−1	−1	1	ROW#3
5	1	−1	1	ROW#1
6	−1	1	1	ROW#2
7	1	1	1	None
SOLN X	*	*	*	Decidable Variables

Note: Asterisk indicates a qualitatively undecidable variable.

Sign pattern information does provide some information about the ranking of entries since positive (resp. negative) entries are larger (resp. smaller) than zero entries. Accordingly, the specification of sgn A and sgn y above include the ranked relationships: $a_{13} > a_{23}$ and $y_1 > y_2$. Suppose that it was known additionally that abs$(a_{11}) > a_{21}$ and abs$(a_{22}) > a_{12}$. For this, the sign pattern of a new row can be constructed by adding row 1 to row 2:

Row	sgn A	sgn y
1 + 2	−1 −1 −1	1

If this row is added to the rows used for the consistency checks, then the solution sign pattern with all positive entries, sgn $x = (1, 1, 1)$, is eliminated. As a result, the additional information enables the determination of A's invertibility and for the problem studied enables a determination of the first column of sgn A^{-1}. An identical analysis finds the second column of sgn A^{-1} all negative for sgn $y = (0, 1, 0)$. An analysis of the system for sgn $y = (0, 0, 1)$ determines the third column of sgn A^{-1} and is given below:

ROW	sgn A			sgn y
1	−1	1	0	0
2	1	−1	−1	0
3	0	1	−1	1
4:1 + 2	−1	−1	−1	0

SOLUTIONS

SOLN#	sgn x Tested			Elimination Row
0	−1	−1	−1	ROW#4
1	1	−1	−1	ROW#1
2	−1	1	−1	ROW#1
3	1	1	−1	None
4	−1	−1	1	ROW#3
5	1	−1	1	ROW#1
6	−1	1	1	ROW#1
7	1	1	1	ROW#4
SOLN X	1	1	−1	Decidable Variables

As a result, the additional information about the outcome of two paired comparisons of the entries of sgn A enabled a complete determination of sgn A^{-1}, as given below:

$$\operatorname{sgn} A^{-1} = \begin{bmatrix} -1 & -1 & 1 \\ -1 & -1 & 1 \\ -1 & -1 & -1 \end{bmatrix}.$$

The information at issue in this example is more than sgn A but less than ord A since all of the entries of A were not ranked. In principle, if all entries were ranked, then new rows to which the elimination principle could be applied for a given solution vector sign pattern could be formed by taking all of the possible sums and differences of rows. As an example, consider the matrix ord A:

ROW	ord A		
1	−1	2	3
2	4	−1	5
3	6	7	−1

The numbers provided for each entry of ord A are to be understood to be (sign-preserved) ordinal values. Accordingly, the analysis of the invertibility of the given ord A is intended to apply equally to any matrix formed from the given matrix by means of an order- and sign-preserving transformation of the particular values given for the entries, i.e., any other matrix with (i) all nonzeros, (ii) an all-negative main diagonal, (iii) all positive off–main diagonal entries, and (iv) the same ranking of the absolute value of the entries. The class of matrices R_A comprises exactly all matrices that can be formed by such a transformation of ord A. For example, the two matrices given below are such transformations of the matrix above. The point of this section is to find conditions for the invertibility of one such matrix that hold for any transformed version of the matrix, i.e., for all members of the class R_A.

ROW	Transformation 1 of ord A		
1	−3	22	23
2	400	−3	500
3	501	700	−3

ROW	Transformation 2 of ord A		
1	−.01	2.2	40
2	40.1	−.01	40.2
3	100	500	−.01

Since all off-diagonal terms are positive, category II.1 cannot hold, since all cycles are positive. Hence, the matrix (or any of its transformations) cannot be a QI-matrix. Given the relative size of the entries, nine additional rows of signs can be formed by taking all possible sums and differences of rows. Additional solution sign patterns (possibly) can be eliminated by one of these additional rows and qualitatively determinable variables detected. The example below utilizes the sign pattern of ord A (and ord y) as the sign patterns of all possible sums and differences of pairs of rows for the case $\text{sgn}(\text{ord } y) = (1, 0, 0)$. Any decidable variables found will correspond to the appropriate entry of the first column of sgn A^{-1}.

ROW	sgn(ord A)			sgn(ord y)
1	−1	1	1	1
2	1	−1	1	0
3	1	1	−1	0
4:1 − 2	−1	1	−1	1
5:2 − 1	1	−1	1	−1
6:1 + 2	1	1	1	1
7:1 − 3	−1	−1	1	1
8:3 − 1	1	1	−1	−1
9:1 + 3	1	1	1	1
10:2 − 3	−1	−1	1	0
11:3 − 2	1	1	−1	0
12:2 + 3	1	1	1	0

SOLUTIONS

SOLN#	sgn x Tested			Elimination Row
0	−1	−1	−1	ROW#6
1	1	−1	−1	ROW#1
2	−1	1	−1	ROW#2
3	1	1	−1	ROW#3
4	−1	−1	1	ROW#3
5	1	−1	1	ROW#2
6	−1	1	1	None
7	1	1	1	ROW#12
SOLN X	−1	1	1	Decidable Variables

The first three rows in the display above are {sgn(ord A), sgn(ord y)}, and the remaining nine are sign patterns resulting from summing or differencing rows in {ord A, ord y} as given (e.g., row 4 is rows 1–2 and row 12 is row 2 + row 3). Then, as before, each of the possible eight (since A is irreducible, zeros do not have to be considered) signed solution vectors is enumerated. For each row in the list of possible solutions, if the solution is eliminated, then the (first) row found for which the solution was inconsistent is indicated. If the solution was consistent for all rows, this is indicated by "None." Inspecting the above results reveals that solutions 0, 6, and 7 are not eliminated by the first three rows; further, each solution variable takes on both a positive and negative sign across the three consistent solutions. As a result (as already noted), there are no qualitatively decidable variables based upon {sgn A, sgn y}. However, given the additional information in ord A, solutions 0 and 7 are eliminated by one of the additional rows. As a result, sgn $x = (-1, 1, 1)$ is the only possible sign pattern in solution to the given system. Since sgn $y = (1, 0, 0)$, the determinate sign pattern found represents the first column of sgn A^{-1}. Further, the additional rows utilized were rows 6 and 12; hence, all of the additional information present in ord A compared to sgn A was used (since only the first row for inconsistency is noted, all of the ranked information may not be used). If the analysis had been repeated for the additional cases, sgn(ord y) = $(0, 1, 0)$ and sgn(ord y) = $(0, 0, 1)$, then all columns of the inverse would have been investigated. For this particular case, the elimination principle enables all entries of the inverse to be signed:

$$\text{sgn}(\text{ord } A^{-1}) = \begin{bmatrix} -1 & 1 & 1 \\ 1 & -1 & 1 \\ 1 & 1 & -1 \end{bmatrix}.$$

An Algebraic Analysis of the Invertibility of ord A

Call a matrix A that can be determined to be invertible based upon ord A an RI-matrix (RI for *ranked invertible*). A feature of the elimination principle as an analytic method is that even if all of the information present in ord A is available, the method does not utilize all of it. In particular, only the relative size of entries in the same column is utilized when constructing new rows from the sums and differences of rows in the original system. Given this, a matrix might be an RI-matrix, but this circumstance is not detectable by utilizing the elimination principle. Consider the matrix ord A and its elimination analysis:[8]

$$\text{ord } A = \begin{bmatrix} -2 & -1 \\ -3 & -4 \end{bmatrix}.$$

ROW	sgn(ord A)		sgn(ord y)
1	-1	-1	1
2	-1	-1	0
$3{:}1-2$	1	1	1
$4{:}2-1$	-1	-1	-1
$5{:}1+2$	-1	-1	1

SOLUTIONS

SOLN#	sgn x Tested		Elimination Row
0	-1	-1	ROW# 2
1	1	-1	None
2	-1	1	None
3	1	1	ROW# 1
SOLN X	*	*	

For the example given, inspection reveals that $\det A > 0$ for the ranking of entries given. As a result, since for a 2×2 matrix, the adjoint is formed by swapping the main diagonal entries and changing the sign (if nonzero) of the off–main diagonal entries, the sign pattern of the adjoint can be readily found, and is the same as the sign pattern of the inverse,

$$\text{sgn}(\text{ord } A^{-1}) = \begin{bmatrix} -1 & 1 \\ 1 & -1 \end{bmatrix}.$$

Thus $\text{sgn } x = (-1, 1)$ for $\text{sgn } y = (1, 0)$. Further, any matrix in the class R_A will have the same sign pattern for its inverse; i.e., the given ord A is an RI-matrix. As shown below, the algebraic method based upon an analysis of cycles in SDG(A) does use all of the information present in ord A and can be used to find the above sign pattern of the inverse matrix; however, the enumeration and evaluation of cycles represents an additional analytic burden.

In order to develop the conditions for A an RI-matrix, it is useful to review the relationship between cycles and the nonzero terms in the expansion of $\det A$. As noted above each nonzero term in the expansion of $\det A$ is the product of n-many nonzero coefficients of A such that each row and each column (respectively, the i and j for a_{ij}) are represented exactly once. If the coefficients are arranged in the natural order of rows, i.e., $a_1 * a_2 * \cdots a_{n*}$, then the sign of the term as it stands is adjusted by multiplying by $(-1)^m$, where m equals the number of times larger column indices precede smaller column indices. Consider further that for any directed arc in SDG(A), and the associated nonzero coefficient of A, the initial vertex of the arc corresponds to a column and the terminal vertex corresponds to a row. For any cycle in SDG(A), each vertex involved appears exactly once as an initial vertex and exactly once as a terminal vertex. As a result of this, among the nonzero terms in the expansion of det A will be terms that correspond to each cycle in SDG(A), where the coefficients in the term correspond to the coefficients that define the cycle, plus main diagonal terms that account for the vertices that are not part of the cycle.

The correspondence between terms in the expansion of det A and cycles can be taken further. Call two cycles *disjoint* if they share no vertices in common. Given this, a term can be found in the expansion of det A that corresponds to any collection of disjoint cycles with main diagonal entries included to account for vertices not present in any of the cycles. It is a result of Maybee's (1966a, 1981) determinantal analysis that every term in the expansion of det A besides the main diagonal term corresponds to a collection of disjoint cycles.[9] Accordingly, the condition for qualitative invertibility set forth in information category II.1 involves the demonstration that any term in the expansion of det A has the sign $(-1)^n$ if all of its embodied cycles are negative (including the main diagonal entries). Call entries in the expansion of det A with sign $(-1)^n$ "friendly" terms, and call terms with the opposite sign "unfriendly" terms (Maybee and Wiener 1988).

Unfriendly terms are terms that embody positive cycles. However, the sign of the product of an even number of all negative cycles is the same as the sign of the product of an even number of all positive cycles. As a result, unfriendly terms are those that embody an odd number of positive cycles. Accordingly, it is necessary to have information about the relative size of entries in addition to their sign amounts and apply it to isolating the unfriendly terms (those with an odd number of positive cycles) and comparing their size to that of the friendly terms in the expansion of det A. For ord A, the comparison of terms must account for the fact that the entries of ord A are only unique up to order-preserving (and sign-preserving) transformations. Let t_g and t_h be the values of the gth and hth terms in the expansion of det A. For each term, list the entries that comprise the term in descending order of absolute value, and let $t_g(a(r))$ and $t_h(a(r))$ be the entries of A in the rth placement. If the finding can be made that

$$\text{abs}(t_g) \geq \text{abs}(t_h) \tag{3.1}$$

for any member of R_A, then it must be that

$$\text{abs}\big(t_g(a(r))\big) \geq \text{abs}(t_h(a(r))) \tag{3.2}$$

for all r. Given this, the strong inequality applies in (3.1) if and only if (3.2) holds for all r and the strong inequality holds for some r. The matrix A is invertible based upon ord A if and only if (i) for every unfriendly term (resp. friendly term) in the expansion of det A, there is a distinct friendly term (resp. unfriendly term) such that (3.1) holds (resp. the negative of (3.1)); and (ii) the strong inequality holds for at least one term. Let FT (resp. UFT) be the index set of friendly (resp. unfriendly) terms in the expansion of det A, and let $n(FT)$ (resp. $n(UFT)$) be the number of indices in the set. The conditions for RI-matrices are as follows.

PROPOSITION 3.1 (RI-MATRIX). *Let A be an irreducible, $n \times n$ matrix in standard form and B be a member of R_A. Then,* $\text{sgn}(\det B) = (-1)^n$ *(resp.* $= -(-1)^n$*) for all B if and only if*

(i) $n(FT) \geq n(UFT)$ *(resp.* $n(UFT) \geq n(FT)$*)*;
(ii) *for every $h \in UFT$ (resp. $h \in FT$), there is a corresponding and distinct $g \in FT$ (resp. $g \in UFT$) such that* $\text{abs}(t_g) \geq \text{abs}(t_h)$; *and*
(iii) *there is at least one $h \in UFT$ (resp. $h \in FT$) with a corresponding and distinct $g \in FT$ (resp. $g \in UFT$) such that* $\text{abs}(t_g) \geq \text{abs}(t_h)$.

It is clear than QI-matrices are also RI-matrices, since QI-matrices have no unfriendly terms in the expansion of det A.[10]

The matrix below is an RI-matrix. Its structure is very simple as there is only one cycle. The term in the expansion of det $A(a_{21}a_{32}a_{13})$ corresponding to the cycle is the single unfriendly term, and the main diagonal term is the single friendly term. The reader can easily confirm that any sign and order preserving transformation of the matrix ($B \in R_A$) is such that sgn det $B < 0$:

$$\text{for ord } A = \begin{bmatrix} -1 & 0 & .5 \\ 1 & -1 & 0 \\ 0 & 1 & -1 \end{bmatrix} \text{ and } \mathbf{B} \in \mathbf{R}_A, \det B < 0.$$

Near-Nonsingularity

An investigation of a matrix to determine its conformance, or nonconformance, with the conditions of Proposition 3.1 presents a substantial algorithmic burden. Generally, the detection of a matrix's satisfaction of these conditions would be facilitated through the addition of some kind of structural restrictions. A compelling first case to consider is that for which the expansion of det A has only one nonzero term of a different sign than the other nonzero terms. Lady, Lundy, and Maybee (1995) termed such matrices "nearly nonsingular," and Lady (1995) termed such matrices "K-matrices." The term *K-matrix* will be used here for ease of expression.

K-matrices can be in two forms. Form A is the form with only one unfriendly term in the expansion of det A, and form B is the form with only one friendly term. From the discussion above, if the matrix is a K-matrix in form A, then there is only one term in the expansion of det A that embodies an odd number of positive cycles. Since every cycle corresponds to its own term in isolation from other cycles, the case at issue then calls for (i) only one positive cycle, which (ii) is not disjoint from any other cycle. A cycle that shares a vertex with every other cycle is termed a *maximal cycle*. Accordingly, for A an irreducible matrix in SF2, A is a K-matrix if it has only one positive cycle and that cycle is a maximal cycle.

For a K-matrix in form B, the nonzero term with the distinct sign is a friendly term. For a matrix in SF2, this term is the term formed by the product of the main diagonal entries. Thus, this second case calls for (i) no negative cycles and, (ii) all maximal positive cycles, to avoid a term that embodies two positive cycles and has the same sign it would have if it embodied two negative cycles (i.e., a friendly term). The results of the discussion can be expressed as follows:

PROPOSITION 3.2 (K-MATRIX) (Lady 1995; Lady, Lundy, and Maybee 1995). *Let A be an $n \times n$ irreducible matrix in $SF2$. A is a K-matrix if and only if for $SDG(A)$,*

(i) *all positive cycles are maximal cycles; and*
(ii) *there is exactly one positive cycle (form A) or all cycles are positive (form B).*

For A a K-matrix, the assessment of invertibility, and the degree to which it can be determined based upon the information in ord A, can be reduced to the analysis of the size of the single term with a distinct sign in comparison to any of the other terms in the expansion of det A. Lady (1995) generalized the K-matrix concept to that of a matrix with more than one unfriendly term, a GK-matrix, but with the value of the sum of the unfriendly terms equal to the value of the sum of the positive cycles, i.e., with all positive cycles maximal, but no restriction on negative cycles. This somewhat more general structure allows the information provided by ord A to be applied to the terms corresponding to the individual positive cycles only, without having to cope with the presence of these cycles in other terms as well (see the discussion below on the invertibility of cycle dominant matrices). Although the conditions for A a QI-matrix, K-matrix, or GK-matrix are restrictive, these cases can be frequently encountered for small (typically sparse) applied systems.

In summary, the analysis to this point provides the following:

Information Category IIIa. The same as category II plus a ranking of the absolute values of the nonzero entries of A.

Class of matrix identified. The matrix of ranked entries (with zeros fixed) ord $A \in R_A$.

Determinable characteristics of A^{-1}

Category IIIa.1 (elimination). For sgn A not a QI-matrix, signs in the kth column of sgn A^{-1} can be detected by applying the elimination principle to all possible sums and differences of the rows of $Ax = y$ for sgn $y_k = 1$ and sgn $y_i = 0$ for $i \neq k$. (Lancaster 1966; Ritschard 1983)

Category IIIa.2 (near-nonsingularity). For A a K-matrix or GK-matrix, the number of unfriendly terms in the expansion of det A is small (only one for the case of a K-matrix). For these special cases, invertibility based upon ord A can be assessed through inspection. (Lady 1995; Lady, Lundy, and Maybee 1995)

The matrix below is a RI-matrix and also a K-matrix in form A:

$$\text{ord } A = \begin{bmatrix} -1 & -1 & 0 & .5 \\ 1 & -1 & -1 & 0 \\ 0 & 1 & -1 & -1 \\ 0 & 0 & 1 & -1 \end{bmatrix}.$$

The reader can confirm that there is a single, positive (i.e., unfriendly) term in the expansion of det A, corresponding to the coefficients, $a_{14}a_{21}a_{32}a_{43}$. The value of this term is .5 for the particular entries selected, but would in any case be less than the value of all of the other (friendly, and equal in value) terms for any sign- and order-preserving transformation of the matrix's entries.

The matrix below is a RI-matrix and also a K-matrix in form B:

$$\text{ord } A = \begin{bmatrix} -.5 & 1 & 1 & 1 \\ 1 & -1 & 0 & 0 \\ 1 & 0 & -1 & 0 \\ 1 & 0 & 0 & -1 \end{bmatrix}.$$

Off-diagonal cycles are positive and involve the vertex 1. Hence, the only friendly term is that corresponding to the main diagonal entries. The value of this term is .5 for the particular entries selected, but would in any case be less than the value of all of the other (unfriendly, and equal in value) terms for any sign- and order-preserving transformation of the matrix's entries.

The matrix below is a RI-matrix and also a GK-matrix:

$$\text{ord}\, A = \begin{bmatrix} -1 & 1 & 1 & 1 \\ .5 & -1 & 0 & 0 \\ .5 & 0 & -1 & 0 \\ -1 & 0 & 0 & -1 \end{bmatrix}.$$

Again, all off-diagonal cycles involve the vertex 1. Two are positive, and one is negative. The terms in the expansion of $\det A$ corresponding to the positive cycles each have the value $-.5$ for the particular entries chosen. The two friendly terms each have the value 1. In any case, the two friendly terms would each be larger than a corresponding unfriendly term for any sign- and order-preserving transformation of the matrix's entries.

Further Restrictions

The centerpiece of the algebraic method for detecting QI-matrices is an analysis of cycles and their signs. For the use of information in addition to sgn A, this strategy will be pursued further. The goal of the analysis is to find cases such that a comparison of the relative size of the value of cycles would be sufficient to determine invertibility. Given this, the analysis becomes a natural extension of the enumeration and signing of cycles undertaken to detect QI-matrices.

To facilitate the analysis, the concept of standard form will be extended. In particular, the point of the analysis will be to isolate positive and negative cycles and reach conclusions about invertibility by comparison of the values of the cycles within the limitations of the information available from a ranking of the values of the matrix entries. Given this, the concept of standard form will be extended to normalize the matrix main diagonal so that the values of terms in the expansion of $\det A$ can be assessed directly with respect to the values of their embodied cycles without concern for the values of main diagonal entries present to account for vertices not included in the cycles. This normalization is accomplished by starting out as before: first, reindexing such that the main diagonal is all nonzero (SF1); and second, multiplying each column by -1 as necessary such that $a_{ii} < 0$ for all i (SF2). The extension is to additionally divide each column by $\text{abs}(a_{ii})$ such that $a_{ii} = -1$ for all i. Define matrices put into this form as being in Standard Form 3 (SF3). It is clear that the invertibility of the matrix put into this form is not influenced by this normalization. However, the transformation may (or may not) disturb the ranked relationships that apply for the entries that appear in different columns. At one convenient extreme, for x an n-vector of solution variables and y an m-vector of parameters, the mathematical system being studied could be in the form

$$f^i(x_j\colon j \neq i; y) - x_i = 0, i = 1,2,\ldots,n.$$

For this case, the Jacobian matrix is presented in SF3 without transformation. The ranked matrix ord A then has the additional information that $a_{ii} = -1$ (and this value can be preserved in its transformations without loss of generality). Alternatively, for the more general case,

$$f^i(x,y) = 0, i = 1,2,\ldots,n,$$

the ability to compare the relative size of entries of the Jacobian matrix across columns may be disturbed by the normalization. For example, the ranked information $a > b > c > d$ does not enable the outcome of the comparison $(a/b)?(c/d)$ to be invariant across increasing monotonic transformations of the values $\{a,b,c,d\}$. As a result, to assume that the entries of A are ranked with the cardinal values 0 and ± 1 fixed points in the ranking is to assume additional information in addition to that provided by ord A. The new class of information being assumed may be specified as follows. For any matrix A in SF3, let $B = [b_{ij}] =$ nord A for all i, be such that[11]

$$\text{sgn } B = \text{sgn } A,$$
$$b_{ij} > b_{gh} \text{ iff } a_{ij} > a_{gh},$$

and

$$b_{ii} = -1.$$

For any matrix A, let NR_A be the class of matrices with entries formed by sign- and order-preserving transformations of A, with entries equal to ± 1 standing as fixed points in the transformations (much as 0 is also a fixed point). If all members of NR_A are invertible, then call nord A an NRI-matrix.

The advantage gained by putting the matrix into SF3 is that a comparison of terms in the expansion of det A can now be conducted through a comparison of the values of the (disjoint) cycles embodied in the terms without regard for the values corresponding to the main diagonal entries. For a matrix A, let c_g and c_h be the gth and hth cycles of SDG(A); let $n1(c*)$, resp. $n2(c*)$, be the number of entries of A corresponding to the arcs of the cycles with values that are no smaller than one, resp. less than one, in absolute value; and let $v(c_g)$ and $v(c_h)$ be the values found by taking the product of the entries corresponding to the arcs of the cycles. The conditions under which conclusions can be reached about the comparative size of the values of cycles based upon the information in nord A are analogous to the conditions given in Proposition 3.1 for comparing entries in the expansion of det A based upon ord A.

For a given cycle, rank the coefficients in the groups counted by $n1(.)$ and $n2(.)$ in the order of descending absolute value such that $c*(a(1r))$ and $c*(a(2r))$ are the entries of a in the rth placement of the ordering.

PROPOSITION 3.3 (COMPARABLE CYCLES). *Let A be an $n \times n$ matrix, let B be a member of NR_A, and let c_g and c_h be the gth and hth cycles of B. Then,* $\text{abs}(v(c_g)) \geq \text{abs}(v(c_h))$ *for all B if and only if*

(i) $n1(c_g) \geq n1(c_h)$, *and* $n2(c_g) \leq n2(c_h)$;
(ii) *for all* $r \leq n1(c_h)$, $c_g(a(1r)) \geq c_h(a(1r))$; *and*
(iii) *for all* $r \leq n2(c_g)$, $c_g(a(2r)) \geq c_h(a(2r))$.

The strong inequality, $\text{abs}(v(c_g)) > \text{abs}(v(c_h))$, holds if the strong inequality holds in either case in (i) or for any r in either (ii) or (iii).

The point of the analysis is to extend the algorithmic principle of the algebraic method for analyzing sgn A, i.e., the identification of cycles and a determination of the sign of each cycle, to an analogous procedure for analyzing nord A, i.e., the identification of cycles and a comparison of their values as provided by Proposition 3.3. Specifically, for a given nord A, let $N = \{1, h\}$ be the index set of negative cycles including the index 1 for the main diagonal term, let $N' = \{h\}$ be the indices of the negative cycles excluding 1, let $P = \{g\}$ be the index set for positive cycles, and assign indices to cycles within each group in descending order of absolute value. The convenience of the K-matrix structure in terms of comparing the values of cycles can be appreciated through inspection of determinantal formulas for this class of matrix:.

A *is a K-matrix in form* A. SDG(A) has only one positive, maximal cycle. The expansion of det A for this case is given by

$$\det A = (-1)^n + (-1)^{n-1}\left[\sum_{g \in N'} v(c_g) + v(c_h)\right]$$
$$+ (\text{other friendly terms}).$$

A *is a K-matrix in form* B. SDG(A) has only positive cycles, and each of these is maximal. The expansion of det A for this case is given by

$$\det A = (-1)^n + (-1)^{n-1}\left[\sum_{h \in P} v(c_h)\right].$$

A K-matrix in form B can be readily put into form A by reindexing to bring any of the positive cycles onto the main diagonal. Let $c(Nr)$ and $c(Pr)$ be the rth cycles. Then the convenient circumstance being sought is expressed by the following,

DEFINITION 3.1 (CYCLE DOMINANCE). *Let A be an $n \times n$, irreducible matrix in $SF3$. A is a cycle dominant matrix if and only if*

(i) $N \geq P$ *(resp.* $P \geq N$*);*
(ii) *for all* $r \in P$ *(resp.* $r \in N$*),* $v(c(Nr)) \geq v(c(Pr))$ *(resp.* $v(c(Pr)) \geq v(c(Nr))$*); and*
(iii) *the negative (resp. positive) cycles are the dominant class if and only if the strong inequality holds for* (i) *and/or some value of r in* (ii).

It is clear that for A in SF3 a cycle dominant matrix, any member of NR_A is also a cycle dominant matrix. The issue then becomes that of the conditions under which cycle dominant matrices can be shown to be invertible.

The Invertibility of Cycle Dominant Matrices

To facilitate the analysis, the matrix will be further transformed. For A a cycle dominant matrix with positive cycles the dominant cycles, reindex the matrix to bring any term in the expansion of det A corresponding to a positive cycle onto the main diagonal. Once done, multiply the columns of the reindexed matrix as necessary such that $a_{ii} = -1$ for all i. Define a cycle dominant matrix in SF3 with negative cycles the dominant class of cycle as being in Standard Form 3* (SF3*). For this class of matrix,

$$\operatorname{sgn}\left((-1) + \sum_{g \in N'} v(c_g) + \sum_{h \in P} v(c_h)\right) = -1.$$

Call this property (CD*). There are a number of conditions on the structure of A that ensure invertibility based upon the property (CD*). The most immediate would the case such that the only terms in the expansion of det A are (as above) the main diagonal term plus terms corresponding to each individual cycle. This circumstance in turn requires that no two cycles be disjoint, i.e., that all cycles be maximal. Matrices with this structure are termed *intercyclic*.

DEFINITION 3.2 (INTERCYCLE). *Let A be an $n \times n$ matrix in $SF1$. A is an intercyclic matrix if and only if all cycles in $SDG(A)$ are maximal cycles.*

PROPOSITION 3.4. *Let A be an $n \times n$ irreducible, intercyclic cycle dominant matrix in $SF3^*$, and let $B \in NR_A$. Then*

$$\operatorname{sgn}\det B = (-1)^n = (-1)^{n-1}\operatorname{sgn}\Bigl((-1) + \sum_{g \in N'} v(c_g) + \sum_{h \in P} v(c_h)\Bigr).$$

PROOF. From the discussion above, only terms in the expansion of $\det A$ with an odd number of positive cycles will have the sign $(-1)^{n-1}$. Since A is intercyclic, the only terms with an odd number of positive cycles are those terms corresponding to each positive cycle individually. Since A is cycle dominant in SF3*, the sum of these terms is smaller than the sum of the terms corresponding to each negative cycle plus the main diagonal term. Accordingly, the size of the sum of the values of the negative cycles dominates. As a result, when the sign convention for the terms is worked out, $\operatorname{sgn}\det A = (-1)^n$.

For the intercyclic structure, there are no terms in the expansion of $\det A$ except those corresponding to individual cycles. This is a very restrictive structural requirement. The basic condition for signing $\det A$ would still hold if there were additional terms if each of these terms could be shown to have the sign $(-1)^n$, i.e., were friendly terms. It is this circumstance that is provided by the concept of a GK-matrix.

DEFINITION 3.3 (GK-MATRIX) (Lady 1995). *Let A be an $n \times n$ irreducible matrix in $SF2$. A is a GK-matrix if and only if*

(i) *no positive cycle in $SDG(A)$ is disjoint from a negative cycle*; *and*
(ii) *no three positive cycles in $SDG(A)$ are disjoint.*

PROPOSITION 3.5. *Let A be an $n \times n$ irreducible matrix in $SF3$. Then*

$$\det A = (-1)^n + (-1)^{n-1}\Bigl[\sum_{g \in N'} v(c_g) + \sum_{h \in P} v(c_h)\Bigr] + \bigl(\text{terms with sgn} = (-1)^n\bigr)$$

if and only if A is a GK-matrix.

PROOF. From the discussion above, the sign of the "other terms" not accounted for in the formula will be $(-1)^{n-1}$ if and only if the number of (disjoint) positive cycles embodied in the term is odd. From (i) of definition 3.3, a GK-matrix has no disjoint positive and negative cycles. As a result, there are no terms that embody exactly one positive cycle except for those accounted for in the formula. Terms with three (or more) positive cycles are excluded from (ii) of definition 3.3. As a result, there are no terms with an odd number of positive cycles except those accounted for in the formula.

COROLLARY. *Let A be an $n \times n$ irreducible, cycle dominant GK-matrix in $SF3^*$ and $B \in NR_A$. Then* $\operatorname{sgn}\det B = (-1)^n$ *(A is an NRI-matrix).*

PROOF. If A is an irreducible, cycle dominant GK-matrix in SF3*, then all members of NR_A have these properties (i.e., these properties are not disturbed by order-preserving transformations of A's entries for which 0 and ± 1 are fixed points). As a result, the property (CD*) holds for all B, and A is an NRI-matrix.

The requirements for a GK-matrix are convenient, but restrictive (intercyclic matrices are GK-matrices). The problem with more general structures and their additional unfriendly terms in the expansion of the determinant that embody more than one cycle is the need to find distinct corresponding friendly terms that are larger, where "larger" is meant in the sense of comparable cycles. For A a cycle dominant matrix in SF3*, for every positive cycle c_h, there exists a negative cycle c_g such that $v(c_g) \geq v(c_h)$, where $v(c_1) = -1$. A simple structural scheme that would accomplish the desired outcome would be the case such that for any unfriendly term in the expansion of $\det A$, the offsetting friendly term could be found by replacing the positive cycles with their corresponding, no smaller, negative cycles. Call structures with this characteristic friendly structures.

Let the vertex set corresponding to the kth cycle be denoted VERT(c_k). Then,

DEFINITION 3.4 (FRIENDLY SDG(A)). *Let A be an $n \times n$ irreducible, cycle dominant matrix in $SF3^*$. $SDG(A)$ is friendly if and only if for c_g a positive cycle that appears in a term in the expansion of $\det A$ that embodies more than one cycle and an odd number of positive cycles, there exists a distinct negative cycle, c_h, such that*

(i) $v(c_h) \geq v(c_g)$; *and*
(ii) VERT(c_h) $\subseteq$ VERT(c_g).

Proposition 3.6. *If A is an $n \times n$ irreducible, cycle dominant matrix in $SF3^*$ with SDG(A) friendly, then A is an NRI-matrix.*

Proof. Given the restrictions on A, any $B \in NR_A$ is such that

$$\operatorname{sgn}\left((-1) + \sum_{g \in N'} v(c_g) + \sum_{h \in P} v(c_h)\right) = -1.$$

Thus, the terms in the expansion of det B that correspond to each cycle individually plus the main diagonal term will have the sign $(-1)^n$. For any term in the expansion of det B that embodies more than one cycle and has the sign $(-1)^{n-1}$, there is a distinct, corresponding term of no smaller value formed by replacing each positive cycle with the corresponding (no smaller) negative cycle. Since SDG(A) is friendly, such offsetting terms can be found for all unfriendly terms. As a result, $\operatorname{sgn}\det B = (-1)^n$ for all $B \in NR_A$.

Although the theorem is true, the algorithmic burden of determining which positive cycles do and which positive cycles do not appear in unfriendly terms that embody more than one cycle is nontrivial. A somewhat more restrictive special case of a matrix with a friendly SDG(A) structure would be that for which every positive cycle could be found to have a "contained" and no smaller negative cycle.

Corollary. *Let A be an $n \times n$ irreducible, cycle dominant matrix in $SF3^*$. If for every positive cycle c_g, there is a distinct negative cycle c_h, $h \in N'$ such that*

(i) $v(c_h) \geq v(c_g)$, *and*
(ii) $\text{VERT}(c_h) \subseteq \text{VERT}(c_g)$,

then A is an NRI-matrix.

Proof. For the conditions given, SDG(A) is clearly friendly and Proposition 3.7 holds.

The matrix below is an NRI-matrix with two positive cycles and a distinct negative cycle (of larger absolute value) embodied within each positive cycle:

$$\text{nord } A = \begin{bmatrix} -1 & -1 & .5 & 0 & 0 & 0 \\ 1 & -1 & 0 & 0 & 0 & 0 \\ 0 & 1 & -1 & 1 & 0 & 0 \\ 0 & 0 & -1 & -1 & -1 & .5 \\ 0 & 0 & 0 & 1 & -1 & 0 \\ 0 & 0 & 0 & 0 & 1 & -1 \end{bmatrix}.$$

Inspection reveals the following structure:

Positive Cycles	Value	Negative Cycles	Value
$1 \to 2 \to 3 \to 1$	.5	$1 \to 2 \to 1$	-1
$4 \to 5 \to 6 \to 4$	.5	$4 \to 5 \to 4$	-1
		$4 \to 3 \to 4$	-1

For every unfriendly term in the expansion of det A there is a distinct and larger (in absolute value) friendly term (as well as the main diagonal term and other friendly terms). Further, this circumstance applies to all $B \in NR_A$.

The additional information (compared to ord A) provided by a normalized main diagonal has significantly increased detectable classes of invertible matrices, as summarized below:

Information Category IIIb. The same as category IIIa plus (re-)resolving the ranking of the absolute value of all nonzero entries of A after dividing each column by the absolute value of its main diagonal entry (which sets the main diagonal entries to -1).

Class of matrix identified. The matrix of ranked entries with normalized main diagonal entries nord $A \in NR_A$.

Determinable characteristics of A^{-1} (*for* A *cycle dominant in* $SF3^*$)

Category IIIb.1 (*GK-matrix*). If A is a GK-matrix, then sgn det $A = -1$. (Lady 1995)

Category IIIb.2 (*SDG*(A) *is friendly*). If SDG(A) is friendly, then sgn det $A = -1$.

3.6 THE INVERTIBILITY OF MATRICES WITH ENTRIES SPECIFIED WITHIN GIVEN INTERVALS

As with the conditions for qualitative determinacy, cycle dominance and associated structural restrictions for an RI-matrix are very restrictive. The next class of information to utilize in assessing a matrix's invertibility is an expression of ranges of values within which the values of the matrix's entries are assumed to lie. Such information would arise naturally in a context for which the values of the entries had been estimated and a given range of error had been associated with each estimate. The point of the analysis here will be to determine a way to use such information in conjunction with an analysis of cycles, as already utilized for QI-matrices, RI-matrices, and NRI-matrices.

For A in any of the standard forms used here, one of the terms in the expansion of det A is the product of the (all nonzero) main diagonal entries. Accordingly, if these entries were sufficiently larger than the matrix's other entries, then invertibility would be ensured, and the sign of this term would dictate the sign of the determinant. The issue becomes, How large is "sufficiently larger"? For a full, quantification of A's entries, one resolution of this issue is the concept of a quasi-dominant diagonal (QDD).

DEFINITION 3.5 (QUASI-DOMINANT DIAGONAL (QDD)). *An $n \times n$ irreducible matrix A has a QDD if there exist $d_j > 0$ such that*

(1) $d_j|a_{jj}| \geq \sum_{i \neq j} d_i|a_{ij}|,\ j = 1, 2, \ldots, n$; *and*

(2) *there is at least one j such that the strong inequality holds.*[12]

It is well known that if A has a QDD, then A is invertible (a QDD also plays an important role in establishing a matrix's stability). The point of the analysis that follows is to bring information about the ranges within which a matrix's entries are assumed to fall as organized by the enumeration and evaluation of cycles into a form that can be related to invertibility. As shown at the end of the section, the development of concepts moves the analysis toward the concept of a matrix with a QDD.

In terms of ranges assumed for the values of a matrix's entries, an upper bound on the value of the sum of positive (and/or negative) cycles is a reasonably straightforward condition to investigate. To develop concepts in parallel to QI-matrices, RI-matrices, and NRI-matrices, assume that for a matrix A, there are two associated matrices, $U = [u_{ij}]$ and $V = [v_{ij}]$, that express (respectively) upper and lower bounds for the absolute value of each entry of A. Assume that the bounds are all nonnegative and sign preserving so that the bounds are such that, for all i and j, and for $a_{ij} \neq 0$,

$$0 < v_{ij} \leq \text{abs}(a_{ij}) \leq u_{ij}.$$

Given this, a matrix "interval class," I_A, could be defined as those matrices with entries that fall within the ranges specified for the entries of A. Specifically,

DEFINITION 3.6 (THE INTERVAL CLASS I_A). *Let A be an $n \times n$ irreducible matrix, and $\{U, V\}$ be nonnegative sign-preserving upper and lower bounds for the entries of A. The interval class of A, I_A, are those $n \times n$ matrices $B = \text{int}\{U, V\}\ A$ such that*

(i) $\text{sgn } B = \text{sgn } A$,

(ii) *if $a_{ij} \neq 0$, then $0 < v_{ij} \leq \text{abs}(b_{ij}) \leq u_{ij}$.*

Given this, interval invertibility may be defined analogously to the definitions for QI-matrices, RI-matrices, and NRI-matrices.

DEFINITION 3.7 (INTERVAL INVERTIBILITY). *Let A be an $n \times n$ irreducible matrix with sign-preserving upper and lower bounds on its entries given by the nonnegative matrices $\{U, V\}$, then A is an II-matrix (interval invertible) if and only if* $\det B \neq 0$ *for all* $B \in I_A$.

From the definition, since $A \in I_A$, A itself must be invertible.

A class of II-matrix can be specified based upon an analysis of the values of positive cycles.

PROPOSITION 3.7 (CYCLE SUMS AND INVERTIBILITY) (Lady 1995). *Let A be an $n \times n$ irreducible matrix in $SF3$ and PCS be the value of the sum of the positive cycles in $SDG(A)$; then if $PCS < 1$,* $\operatorname{sgn}\det A = (-1)^n$.

This is a particularly useful result since the conditions given do not require any additional restrictions on A's structure. The result can be generalized further to be independent of A's (off-diagonal) sign pattern as well.

DEFINITION 3.7 (SLO-MATRIX FOR SUM LESS THAN ONE) (Lady 1996). *Let A be an $n \times n$ irreducible matrix in $SF3$, PCS be the value of the sum of positive cycles, and NCS be the sum of the values of the negative cycles of $SDG(A)$; then A is a SLO-matrix if and only if $PCS - NCS < 1$.*

PROPOSITION 3.8 (EQUIVALENCE OF FORMS) (Lady 1996). *Let A be an $n \times n$ irreducible matrix in $SF3$; then*

(i) *if A is a SLO-matrix, then A has a QDD; and*
(ii) *if A is intercyclic, then A is a SLO-matrix if and only if A has a QDD.*

These results can be immediately combined for the purposes of this section. For a given matrix and bounds $\{A, U, V\}$, let $B_{\max}$ be a matrix in SF3 formed as follows: first, setting all off-diagonal entries of A at the upper bound on their absolute value; second, setting all main diagonal entries at the lower bound on their absolute value; and third, putting the array into SF3 by dividing each column by its main diagonal entry. Given this, the off-diagonal terms are at their largest absolute values for any member of I_A put into SF3.

PROPOSITION 3.9 (A CLASS OF II-MATRIX). *Let A be an $n \times n$ matrix irreducible matrix with sign-preserving bounds $\{U, V\}$, $B_{\max} \in I_A$ in $SF3$ as described above, and PCS equal to the sum of the values of positive cycles in $SDG(B_{\max})$; then if $PCS < 1$, A is an II-matrix.*

PROOF. B_{max} was constructed such that the value of the sum of its positive cycles is not exceeded for any other member of I_A. As a result, $PCS < 1$ for all members of I_A. From Proposition 3.7, under these circumstances, $\operatorname{sgn}\det B = -1$ for all $B \in I_A$, and A is an II-matrix.

COROLLARY. *Let B_{max} be defined as above. If B_{max} is a SLO-matrix, then all members of I_A are SLO-matrices.*

Lady (1996) showed that SLO-matrices were not only invertible but also stable and Hicksian stable matrices. The route of derivation is particularly convenient in that the valuation of cycles can arise naturally from algorithmic principles designed to enumerate cycles and evaluate their signs by evaluating the sign of the product of the matrix entries that correspond to the arcs in a cycle. Otherwise, preprocessing a matrix with bounded elements into the B_{max} from is straightforward.

These results provide the following:

Information Category IV. The same as category IIIa plus the specification of sign-preserving upper and lower bounds on the absolute values of the nonzero entries of A.

Class of matrix identified. The matrix of bounded entries (with zeros fixed) $\operatorname{int}\{U, V\}\ A \in I_A$.

Determinable characteristics of A^{-1}

*Category IV.*1. For B_{max} in SF3, all members of I_A are invertible if the sum of the positive cycles in SDG(B_{max}) is less than one. (Derived from Lady 1995)

*Category IV.*2 (*SLO-matrices*). For B_{max} in SF3, all members of I_A are invertible for any sign pattern of the off-diagonal elements if the sum of the absolute values of all cycles in SDG(B_{max}) is less than one. (Lady 1996)

Let

$$\operatorname{int}\{U, V\}A = \begin{bmatrix} -1 & .3 & .3 \\ .3 & -1 & .3 \\ .3 & .3 & -1 \end{bmatrix},$$

with $U = 1.2A$ and $V = .8A$. Then

$$B_{max} = \begin{bmatrix} -.8 & .36 & .36 \\ .36 & -.8 & .36 \\ .36 & .36 & -.8 \end{bmatrix},$$

and for $C = B_{\max}$ in SF3,

$$C = \begin{bmatrix} -1 & .45 & .45 \\ .45 & -1 & .45 \\ .45 & .45 & -1 \end{bmatrix}.$$

Since all off-diagonal entries are positive, all cycles have a positive value:

Cycles	Cycle Values
$1 \to 2 \to 1$	.202
$1 \to 3 \to 1$	.202
$2 \to 3 \to 2$	.202
$1 \to 2 \to 3 \to 1$	.091
$1 \to 3 \to 2 \to 1$	.091

Sum of positive cycle values = .788.

As a result, $B_{\max}$ is a SLO-matrix. Further, any matrix with entries within the ranges $\text{int}\{U, V\}A$ will also have the sum of its positive cycles less than unity when put into SF3. As a result, $\text{int}\{U, V\}A$ is an II-matrix.

3.7 SUMMARY OF THE ALGEBRAIC METHOD

Although the elimination principle provides a straightforward approach to investigating invertibility on the basis of sign pattern and some ranking information, the approach suffers from problems of efficiency and has not been shown to enable a processing of interval ranges for a matrix's entries. Alternatively, the algebraic approach could be used to find classes of invertible matrices for each category of information to which it was applied. The core of this approach is the investigation of cycles, the signs of cycles, and the value of cycles in the signed directed graph corresponding to a matrix put into a standard form. The approach taken here for processing information in addition to a matrix sign pattern was to use this information in a comparison of the size of individual cycles. The restrictions provided for the several cases developed had the effect of determining the outcome of comparisons (of the values of terms in the expansion of the determinant) that were not actually made, on the basis of the comparisons of the values of the individual cycles. Given this, the findings are generally sufficient. The advantage of the approach taken is that it builds directly upon an

TABLE 3.1
Information Categories and Invertibility

Information Category I. bol $A \in B_A$: Specification of zero and nonzero entries, i.e., which variables appear in which equations. **Conditions for Invertibility.** *Necessary.* BI-matrix: All members of B_A are invertible only if A can be put into SF1.

Information Category II. sgn $A \in Q_A$: Specification of the signs of the nonzero entries of A. **Conditions for Invertibility.** *Necessary and sufficient.* QI-matrix: For A in SF2, all members of Q_A are invertible if and only if the signs of all cycles in SDG(A) are negative.

Information Category IIIa. ord $A \in R_A$: Specification of a ranking of the absolute values of the nonzero entries of A. **Conditions for Invertibility.** *Sufficient.* RI-matrix: All members of R_A are invertible if any solution variable can be signed by using the elimination principle. For ord A a K-matrix in SF2, all members of R_A are invertible if the distinctly signed term in the expansion of det A is comparable to and smaller than another term.

Information Category IIIb. nord $A \in NR_A$: Specification of a ranking of the absolute values of the nonzero off-diagonal entries after setting the on-diagonal entries $= -1$. **Conditions for Invertibility.** *Sufficient.* NRI-matrix: For nord A a cycle dominant matrix in SF*, all members of NR_A are invertible if nord A is a GK-matrix or if SDG(A) is friendly.

Information Category IV. int$\{U, V\}A \in I_A$: Specification of sign-preserving upper and lower bounds on the absolute value of A's nonzero entries. **Conditions for Invertibility.** *Sufficient.* II-matrix: For $B_{\max} \in I_A$ in SF3, all members of I_A are invertible if the sum of the values of the positive cycles in SDG($B_{\max}$) is less than one.

Note: All matrices are assumed to be irreducible.

analysis of cycles for the qualitative case (category II). Accordingly, algorithmic approaches that implement the inspection of the signs of cycles could be readily amended to process information on the values of cycles. Given this, it would be possible in principle to bring an investigation of all of the information categories into a unified algorithmic regime. A summary of the results for invertibility derived in this chapter is provided in Table 3.1.

3.8 STABILITY

A matrix's stability will be assessed with respect to two criteria.[13] The matrix A is a *Hicksian stable matrix* if and only if odd-order principal minors are negative and even-order principal minors are positive. The matrix A is a *stable matrix* if the real parts of the characteristic roots of A are negative. The analyses above generally concern conditions for an $n \times n$ irreducible matrix such that it can be shown that sgn det $A = (-1)^n$ on the basis of the information assumed to be available. Such a matrix (in the appropriate standard form) would be Hicksian stable if it could be found that if any k rows and columns (of the same index) were removed, $k = 1, 2, \ldots, n-1$, then the residual $m \times m$ array A' (for $m = n - k$) also necessarily would be such that sgn det $A' = (-1)^m$. A QI-matrix is an example of the principle at issue. For sgn A (in SF2) such that all cycles are negative in SDG(A), if k-many rows and columns of the same index are removed, the residual array A' is also a QI-matrix since SDG(A') cannot have positive cycles if SDG(A) does not; i.e., removing vertices could eliminate existing cycles, but cannot change the sign of an existing cycle or introduce a new cycle. Accordingly, if sgn A is a QI-matrix, then sgn A is a Hicksian (sign) stable matrix. From Theorem 2.3, if A is a QI-matrix, all members of Q_A are also stable matrices if and only if SGD(A) has no cycles that involve more than two vertices.

The discussion of stability in this section will be limited to the information categories summarized in Table 3.1. For information category I, since invertibility is a necessary condition for stability, a matrix is a Hicksian stable, or stable, matrix only if bol A can be put into SF1. For RI-matrices, the general approach for establishing invertibility is to find, for each member of the group of terms in the expansion of det A of a given sign (say, unfriendly terms), that there is a distinct term of the opposite sign that is "larger" in the sense of comparable terms. Such a finding might not survive as a characteristic of all of the residual arrays, A', formed by deleting rows and columns of ord A, since some of the "larger" terms might be eliminated but the terms to which they correspond not eliminated. Manipulatively, the issue is expressed by the fact that the removal of vertices in SDG(A) could remove cycles for friendly terms but not for the corresponding unfriendly terms. For a GK-matrix (say) in form A, the issue could be resolved for Hicksian stability if there were a friendly term in the expansion of det A that could be brought onto the main diagonal such that the entry of minimum absolute value in the term were larger than the entry with the maximum absolute value in the unfriendly term. Given this, as the residual arrays are formed, the main diagonal term in the expansion of det A' would remain larger (in absolute value) than the single term with the opposite sign, assuming the vertices defining the cycle remained. This result is expressed in Proposition 3.10.

PROPOSITION 3.10. *Let* ord A *be an* $n \times n$ *irreducible K-matrix in form A,* $E(P)$ *be the entries of* ord A *corresponding to the single, positive cycle,*

a_{max} be the (an) entry of ord *A belonging to E(P) with the largest absolute value, and aa the main diagonal entry of* ord *A with the smallest absolute value; then, if* abs(*aa*) > abs(a_{max}), *all members of* R_A *are Hicksian stable.*

PROOF. All terms in the expansion of det ord A have the sign $(-1)^n$ except the term corresponding to the single positive cycle. This (unfriendly) term, with value $v(UFT)$, is the product of the entries $E(P)$ plus main diagonal entries corresponding to vertices not accounted for by the entries $E(P)$. From the assumptions given, the value of the main diagonal term, with value MD, is larger, $MD > v(UFT)$. Further, as subarrays are formed with corresponding terms in the expansion of det A', $v(UFT')$, and MD', it remains true that $MD' > v(UFT')$ for all ord A'. This inequality survives any sign- and order-preserving transformation of ord A'; hence, all members of R_A are Hicksian stable.

The conditions for true stability are more difficult to approach. The results here will be based upon the following two important results.

THEOREM 3.1 (CONDITIONS FOR A STABLE) (McKenzie 1960a). *For A an $n \times n$ (irreducible) matrix in SF2, if A has a QDD, then A is a stable matrix.*

THEOREM 3.2 (EQUIVALENCE OF STABILITY CONDITIONS) (Metzler 1945). *For A an $n \times n$ (irreducible) matrix in SF2, if $a_{ij} \geq 0$ for $i \neq j$, then the following are equivalent:*

(1) *A is a stable matrix*
(2) *A is a Hicksian stable matrix*

From the discussion leading up to Proposition 3.10, if SDG(A) contains only one cycle, and that cycle has a positive sign, then for A irreducible, $E(P)$ must account for all of the vertices of SDG(A). As a result, for any of the arrays A', the main diagonal term in the expansion of det A' is the only nonzero term. It is clear for this extremely restrictive case that the conditions for A to have a QDD would be satisfied since each row and column would have exactly one off-diagonal entry, and each main diagonal entry would be larger (in absolute value) than any of these. This result is stated formally.

PROPOSITION 3.11. *Let* ord *A be an $n \times n$ irreducible K-matrix in both form A and form B and aa the main diagonal entry of a with the minimum absolute value. If* abs(*aa*) > abs(a_{ij}) *for $i \neq j$, then all members of R_A are* (i) *Hicksian stable matrices and* (ii) *stable matrices.*

PROOF. The conditions given satisfy the conditions for Proposition 3.10, establishing (i). Since each row and column have only one off-diagonal term, ord A has a QDD; given this, (ii) follows from Theorem 3.2.

The matrix below satisfies the conditions of Proposition 3.11:

$$\text{ord } A = \begin{bmatrix} -1 & 0 & 0 & .5 \\ .5 & -1 & 0 & 0 \\ 0 & .5 & -1 & 0 \\ 0 & 0 & .5 & -1 \end{bmatrix}.$$

It is clear that if the conditions of the proposition are satisfied, the results are otherwise independent of the signs of the nonzero off-diagonal entries (e.g., for this case, all off-diagonal entries could be negative, or two could be negative, and the conditions of the proposition would still be satisfied).

For matrices with ranked entries and a normalized main diagonal, the analysis can focus specifically upon the "values" of cycles without further accounting for the diagonal entries. In this frame of reference for nord A a cycle dominant matrix in SF*, the problem mentioned above remains with respect to principal minors of nord A: the formation of the subarrays could eliminate negative cycles but not the positive cycles they were offsetting in evaluating th sign of det nord A. On the other hand, the signs of the principal minors can be determined if SDG(A) is friendly. For this case, each positive cycle is offset by a distinct, larger (in absolute value) negative cycle with indices embodied within the index set of the positive cycle. If the negative cycle is eliminated for a subarray, then the corresponding positive cycle is also eliminated. Accordingly, such matrices are Hicksian stable.

PROPOSITION 3.12. *Let* nord A *be a cycle dominant matrix in SF*. If SDG(A) is friendly, then all members of NR_A are Hicksian stable.*

PROOF. For any $m \times m$ subarray, A', if a positive cycle remains, its offsetting negative cycle also remains. As a result, for any $m \times m$ principal minor A', $\operatorname{sgn}\det A' = (-1)^m$. As a result, nord A is Hicksian stable, as are any sign-, -1- and order-preserving transformations of nord A.

The following matrix satisfies the conditions of Proposition 3.12:

$$\text{nord } A = \begin{bmatrix} -1 & -3 & 0 & 2 \\ 1 & -1 & 0 & 0 \\ 0 & 1 & -1 & 0 \\ 0 & 0 & 1 & -1 \end{bmatrix}.$$

There is one positive cycle involving all of the vertices and one negative cycle as given below:

Cycle	Value
$1 \to 2 \to 3 \to 4 \to 1$	2
$1 \to 2 \to 1$	-3

For the conditions available from the information categories being utilized here, if there is more than one positive cycle and/or negative cycles involving more than two vertices, cases for which all members of a given NR_A are stable matrices have not been identified.

For interval classes of matrices, results from Proposition 3.9 and its corollary can be related directly to stability criteria. From Proposition 3.9, consider the class I_A corresponding to the matrix int $A\{U,V\}$ such that the sum of the positive cycles, PCS, in the array $B_{\max}$ satisfy $PCS < 1$. For any of the subarrays $B'_{\max}$, the corresponding sum $PCS' < 1$ as well, since deleting vertices cannot increase the value of the sum of the remaining positive cycles. Given this, $B_{\max}$ is Hicksian stable. If $B_{\max}$ is a SLO-matrix, then $B_{\max}$ is Hicksian stable for any sign pattern of its nonzero, off-diagonal entries, including the case of nonnegative off-diagonal entries as called for in Theorem 3.2. Given this, if $B_{\max}$ is a SLO-matrix, then $B_{\max}$ is a stable matrix. These results are stated formally.

THEOREM 3.3 (Lady 1996). *Let A be an $n \times n$ irreducible matrix with sign-preserving bounds $\{U,V\}$, $B_{\max} \in I_A$ in $SF3$ as described above, PCS the sum of the values of positive cycles, and NCS the sum of the values of negative cycles in $SDG(B_{\max})$; then*

(1) *if $PCS < 1$, all members of I_A are Hicksian stable matrices*; *and*
(2) *if $PCS - NCS < 1$, all members of I_A are stable matrices.*

Let the matrix below be $B_{\max}$ for a given int $A\{U,V\}$:

$$B_{\max} = \begin{bmatrix} -1 & .5 & 0 & 0 \\ -.5 & -1 & -.5 & 0 \\ 0 & -.5 & -1 & .5 \\ 0 & 0 & .5 & -1 \end{bmatrix}.$$

An analysis of cycles is as follows:

Cycle	Value
$1 \to 2 \to 1$	$-.25$
$2 \to 3 \to 2$	.25
$3 \to 4 \to 3$	.25

Given this, $PCS = .5$ and $NCS = -.25$; thus, $PCS - NCS = .75$, which is < 1. As a result, $B_{\max}$ is a SLO-matrix and all members of I_A are both Hicksian stable and stable matrices.

As with the analogous results for invertibility, most of the conditions provided in this section are sufficient conditions. As a result, there remains the potential for expanding upon the cases for which less than a full quantification of the entries of the Jacobian matrix can be shown to establish stability or Hicksian stability. The results given in this section for stability are summarized in Table 3.2.

TABLE 3.2
Information Categories and Stability

Information Category I. bol $A \in B_A$: Specification of zero and nonzero entries, i.e., which variables appear in which equations. **Conditions for A Stable and Hicksian Stable.** *Necessary.* BI-matrix: All members of B_A are invertible only if A can be put into SF1.

Information Category II. sgn $A \in Q_A$: Specification of the signs of the nonzero entries of A. **Conditions for Hicksian Stability.** *Necessary and sufficient.* QI-matrix: For A in SF2, all members of Q_A are Hicksian stable if and only if the signs of all cycles in SDG(A) are negative. **Conditions for Stability.** *Necessary and sufficient.* All members of Q_A are stable matrices if and only if A is a QI-matrix with no cycles involving more than two vertices.

Information Category IIIa. ord $A \in R_A$: Specification of a ranking of the absolute values of the nonzero entries of A. **Conditions for Hicksian Stability.** *Sufficient.* All members of R_A are Hicksian stable for ord A a K-matrix in form A and abs(aa) > abs($a_{\max}$), as given in Proposition 3.10.

Information Category IIIb. nord $A \in NR_A$: Specification of a ranking of the absolute values of the nonzero off-diagonal entries after setting the on-diagonal entries $= -1$. **Conditions for A Stable and Hicksian Stable.** *Sufficient.* For nord$A$ a K-matrix in forms A and B with abs(aa) > abs($a_{\max}$) as above, then all members of NR_A are stable and Hicksian stable. **Conditions for Hicksian Stability.** *Sufficient.* For nord A a cycle dominant matrix in SF*, if SDG(A) is friendly, then all members of NR_A are Hicksian stable.

Information Category IV. int$\{U, V\}A \in I_A$: Specification of sign-preserving upper and lower bounds on the absolute value of A's nonzero entries. **Conditions for Stability and Hicksian Stability.** *Sufficient.* For $B_{\max} \in I_A$ in SF3, all members of I_A are Hicksian stable if $PCS < 1$ and all members are stable matrices if $PCS - NCS < 1$ ($B_{\max}$ is a SLO-matrix).

Note: All matrices are assumed to be irreducible.

4

Applications in Qualitative Comparative Statics

4.1 INTRODUCTION

In this chapter, three rather different models are investigated with respect to their qualitative comparative statics: the Oil Market Simulation and Oil and Gas Supply models, both developed by the U.S. Department of Energy, and a small macro-economic model sometimes called Klein's Model I. In each case, the model may be understood to be a system of equations,

$$f^i(x, y) = 0, i = 1, 2, \dots, n, \tag{4.1}$$

where x is an n-vector of variables to be evaluated by solving the model and y is an m-vector of parameters to be given values prior to solving the model. The comparative statics analysis of a model solution is conducted in terms of the linear system

$$\sum_{j=1}^{n} \frac{\partial f^i}{\partial x_j} dx_j = - \sum_{k=1}^{m} \frac{\partial f^i}{\partial y_k} dy_k, i = 1, 2, \dots, n, \tag{4.2}$$

where the partial derivatives in (4.2) are evaluated at the reference solution. As above, notation may be simplified to that of a simple linear system, $Ax = b$, such that for

$$A = \left[\frac{\partial f^i}{\partial x_j} \right],$$

b and x can be written as

$$b = \left[- \sum_{k=1}^{m} \frac{\partial f^i}{\partial y_k} dy_k \right],$$

and

$$x = [dx_j], i, j = 1, 2, \ldots, n.$$

Qualitative comparative statics is then the problem: what can be said about the signs of the entries of A^{-1}, given a specification of the signs of the entries of A; and given this, what can be said about the signs of the entries of x, given the signs of the entries of b. If some (or all) of the signs at issue cannot be found by means of a purely qualitative analysis, then other information might also be used, such as the types of information present in the various information categories discussed in Chapter 3. Generically, such additional information is sometimes called *side conditions*.

There is a wide range of motivation for the conduct of such an analysis. At one extreme is the issue of the "scientific" content of theory. If theoretical propositions require only that the entries of A have certain signs, then, strictly speaking, the theory cannot be tested, i.e., falsified (Popper [1934] 1959), unless just a knowledge of the signs in A can be shown to constrain (at least some of) the signs in A^{-1}. At another extreme, necessary correspondences between the signs of A and its inverse might be used to audit an actual model installed on a computer (Hale and Lady 1995). The qualitative results could be used to specify experiments with the computer model. If the necessary directions of influence were not found between parameters (i.e., model assumptions) and (solution) variables for the computer version of the model, then the model's written documentation would be found to be inadequate.

Perhaps the strongest motivation for qualitative analysis can be found in the general enterprise of using conceptual models at all. Across many subject matters, the organization of phenomena through systems of relationships does not readily, if at all, submit to a full quantification of the relationships. Even when it does, and economics provides many a good example of this, the actual magnitudes found must be accepted as error prone at best, and usually transitory in principle. Conclusions with the conceptual model might not be possible unless, or at least would be substantially more robust if, the conclusions could be based upon something less than a full quantification of the conceptual framework.

4.2 ALGORITHMIC PRINCIPLES

The elimination method will not be used in this section. Instead, the method used will be algebraic: putting A into a required standard form (SF2, as described in Chapter 3) and conducting tests to determine if the signs of cycles in the corresponding signed digraph satisfy certain conditions. If they do, then a variety of mathematical results are available that specify what the qualitatively decidable properties of the system are. This is an efficient method, particularly for large sparse systems; further, if side conditions are added to the analysis, this approach must necessarily detect any conclusions that can be found with the side conditions. However, there are a number of problems in applying this approach, including those of putting the A matrix into the required standard form and detecting and working with any maximal irreducible substructures within the digraph (i.e., strong components). These problems are by no means trivial and the analysis is considerably facilitated through computer support.[1]

Finding and Signing Cycles

The algorithm used to study the examples provided below was designed to detect and sign cycles of inference associated with the signed digraph corresponding to sgn A. Before the algorithm can be applied, the A-matrix must be studied to identify its strong components; the (irreducible) submatrices corresponding to each strong component must be isolated for individual analysis; and given this, each submatrix must be put into a standard form prior to further analysis. For large systems with a rich pattern of maximal substructure interrelationships, all of this preprocessing can represent a substantial burden.[2] On the other hand, for systems of a relatively modest size, even with tens of variables, a qualitative analysis based on identifying and signing cycles often can be readily undertaken with pencil and paper. For example, the Oil Market Simulation Model (OMS) discussed in the next section has a 19×19, irreducible Jacobian matrix. An analysis of the model's inference structure revealed only fifteen cycles based upon a repetitive pattern of variable interdependence. The enumeration and signing of these cycles, and the consequent signing of the entries of the Jacobian inverse could be done by hand (although in fact these tasks were done here by computer).

For the purposes of discussion here, assume the matrix A is irreducible. The conduct of a qualitative analysis based on cycles and their signs proceeds as follows:

First, find the decidable signs in sgn A^{-1}, if any, as follows:

Step 1. Put A into standard form SF2 by (1.1) renumbering variables as needed such that $a_{ii} \neq 0$ for all i; and (1.2) multiplying each column in the reindexed matrix by -1 as necessary such that $a_{ii} < 0$ for all i.

Step 2. Enumerate all cycles of inference in the signed digraph corresponding to the matrix when in standard form. Compute the sign of each cycle as the sign of the product of the coefficients of A that define the cycle.

Step 3. If any cycle is positive, then stop: sgn A^{-1} has no qualitatively decidable entries. If all cycles are negative, then call sgn A *qualitatively invertible*, i.e., a QI-matrix, and continue.

Step 4. For sgn A a QI-matrix in SF2, if sgn $a_{ij} \neq 0$, then sgn $a_{ji}^{-1} =$ sgn a_{ij}.

Step 5a. As might be done by computer, if sgn $a_{ij} = 0$, then form the submatrix corresponding to the (i, j)th cofactor of A. Repeat steps 1 and 2 above. In performing step 1, keep track of the consequences for the sign of the determinant of the transformed array due to the column swapping and sign changing. If any cycle is positive, then stop: sgn a_{ji}^{-1} is qualitatively indeterminate. If all cycles are negative, then sgn $a_{ji}^{-1} = (-1)^n$ sgn(i, j)th cofactor).

Step 5b. If one is using pencil and paper, it is easier to apply the theorem of Maybee and Quirk (1969). If sgn $a_{ij} = 0$, then evaluate all of the signs of the paths $p(i \rightarrow j)$. If any two paths have different signs, then stop: sgn a_{ji}^{-1} is qualitatively indeterminate. If all paths have the same sign, then sgn $a_{ji}^{-1} = -$sgn $p(i \rightarrow j)$.

Now, for sgn A qualitatively invertible, find the decidable signs in sgn x:

Step 6. Consider the array of signs, [sgn A^{-1} sgn b]. If the nonzero signs in any row are all of the same sign, the corresponding entry of sgn x is qualitatively determinate and equal to the given sign. If a row has both positive and negative signs, the corresponding entry of sgn x is not qualitatively determinate.

As an example, consider the Lagrangian function corresponding to a two-factor, cost-constrained, output maximization problem,

$$L = f(x_1, x_2) - \lambda[w_1 x_1 + w_2 x_2 - TC],$$

where $f(\ )$ is the two-factor production function, x is a 2-vector of input quantities, w is a 2-vector of input prices, and TC is the amount of total cost constraining input choices. As described in Chapter 5, comparative statics analyses of systems such as this one can be conducted under the maximization hypothesis (a side condition). But instead, assume that only the sign pattern of the bordered Hessian matrix, sgn BH, for this problem is known (hence, among other things, satisfaction of the second-order conditions for a solution to the constrained maximization problem remains an issue).

What signs should be assumed for the entries of sgn BH? From the structure of the functions involved and the assumption of positive factor prices, the (3,3) entry is zero, and the other entries of the third row and column are negative. The signs to assume, then, are the signs in the Hessian of the production function, $F = [f_{ij}]$. Assume, as special case, that the on-diagonal signs of F are negative and the off-diagonal signs (which are sign symmetric) are positive. Given this, the sign pattern of sgn BH is

$$\text{sgn}(BH) = \begin{bmatrix} -1 & 1 & -1 \\ 1 & -1 & -1 \\ -1 & -1 & 0 \end{bmatrix}.$$

In invoking the algorithm outlined above, the first step is to construct a transformed matrix such that the main diagonal entries are all negative. Inspection reveals that transformation of sgn BH into sgn $BH^* =$ sgn A by swapping the second and third columns (i.e., making λ variable #2 and x_2 variable #3) results in a matrix in the standard form for further analysis:

$$\text{sgn}(BH^*) = \text{sgn}(A) = \begin{bmatrix} -1 & -1 & 1 \\ 1 & -1 & -1 \\ -1 & 0 & -1 \end{bmatrix}.$$

The second step is to enumerate and sign all cycles of inference in the signed digraph corresponding to sgn A. This can easily be done by hand for this small system and the results are as follows:

Cycle	*Sign of Cycle*
$a_{12}(+)a_{21}(-)$	-1
$a_{13}(+)a_{31}(-)$	-1
$a_{31}(-)a_{23}(-)a_{12}(-)$	-1

No cycle is positive, and as a result the analysis can continue. From step 4, all but one entry in sgn A^{-1} can be determined. Since $a_{32} = 0$, sgn a_{23}^{-1} remains in doubt. Applying step 5a by inspecting the subarray for the (3,2) cofactor (which does not have to be transformed) reveals a positive cycle, so sgn a_{23}^{-1} is undecidable. Applying step 5b reveals that there are two paths $p(3 \to 2)$ with opposite signs, sustaining the result found by applying step 5a. Given this, all but one sign in sgn A^{-1} can be determined. This result is

$$\text{sgn } A^{-1} = \begin{bmatrix} -1 & 1 & -1 \\ -1 & -1 & * \\ 1 & -1 & -1 \end{bmatrix}.$$

Transforming this array back to account for the manipulation of sgn BH into a standard from entails swapping the second and third rows. Assuming $\lambda > 0$ in solution to the constrained maximization problem, the resulting array, [sgn BH^{-1} sgn b], based upon a qualitative analysis (and, for sgn b, a selected "right side" for the comparative statics problem), is

$$[\text{sgn}(BH^{-1})\text{sgn}(b)] = \begin{bmatrix} \frac{dx_1^*}{dw_1}(-) & \frac{dx_1^*}{dw_2}(+) & -\frac{dx_1}{dTC}(-) \\ \frac{dx_2^*}{dw_1}(+) & \frac{dx_2^*}{dw_2}(-) & -\frac{dx_2}{dTC}(-) \\ \frac{d\lambda^*}{dw_1}(-) & \frac{d\lambda^*}{dw_2}(-) & -\frac{d\lambda}{dTC}(?) \end{bmatrix},$$

where the asterisk superscript denotes changes such that production is held constant (i.e., a substitution effect). Since sgn A is qualitatively invertible, odd-order principal minors are negative, and even order positive, which are the second-order conditions for the constrained maximization problem as expressed for the bordered Hessian transformed here into a standard form.

Limitations of Method

The qualitative analysis works well for this small example, but would not be sustained for larger, constrained optimization problems. The reason for this is that the Hessian for the production (or utility) function is sign symmetric. This symmetry was eliminated for the two-input case when transforming the bordered Hessian into a standard form. For constrained optimization problems with more than two substantive variables (plus the Lagrangian), the (signed directed graph corresponding to the) bordered Hessian matrix (when put into SF2) would embody at least one positive cycle, if the Hessian had any off-diagonal nonzeros. Thus side conditions derived from the maximization hypothesis are required to resolve the signs of entries in the inverse bordered (or unbordered) Hessian (this explains the somewhat different methological approach in Chapter 5). As a result, the examples provided below are for models of multi-market equilibria or other classes of the nonoptimization problem.

4.3 A QUALITATIVE ANALYSIS OF THE OIL MARKET SIMULATION MODEL

The Oil Market Simulation Model (OMS) was part of the energy-forecasting capabilities maintained by the Energy Information Administration (EIA) of the U.S. Department of Energy. The OMS solves for international energy market equilibrium as an analytic precursor to the detailed forecast of U.S. domestic energy markets. The essential purpose of the OMS is to provide a basis for the supply to the United States of imported oil and petroleum products in a fashion that accounts for worldwide petroleum supply and demand. The OMS was used in conjunction with other forecasting tools by the EIA in the preparation of energy systems forecasts in the period 1985–1993. The system has evolved since that time away from the use of OMS and many associated models and is currently the National Energy Modeling System (NEMS). The 1990 version of OMS is used for this example.[3]

In summary, the OMS partitions world oil markets into seven consuming regions and nine producing regions. OPEC is one of the producing regions. The demand for OPEC oil is found as the residual of regional demand less regional supply. The world oil price is found as a function of OPEC's capacity utilization, as determined by a comparison of OPEC production to the demand for OPEC oil. As formulated, the model is fully simultaneous and comprises nineteen equations. Each equation, or group of equations, is summarized below and manipulated into the form consistent with (4.2) above. As necessary, equations and variables are indexed such that the main diagonal terms of the resulting Jacobian matrix are all nonzero.

The Equations

Regional Demand Equations

The demand relationship for the ith region, $i = 1, 2, \ldots, 7$, has the form

$$D_i = Y_i^{gi} * P^{di} * D_{i-}^{ci},$$

where D_i = oil demand in demand region i, Y_i = income in region i, P = world oil price, D_{i-} = demand in region i lagged one period, gi = the income elasticity of demand in i, di = the price elasticity of demand in i, and ci = a speed of dynamic adjustment factor in i. Prior to solution to the model, a GDP price feedback is given as, $Y_i = P^{fi}$, where fi = the feedback elasticity. Making this substitution into the demand equations here yields

$$D_i = P^{ei} * D_{i-}^{ci}, i = 1, 2, \ldots, 7,$$

where $ei = (fi * gi) + di$. Inspection of model data inputs shows the gi to be positive and no larger than 1, the di negative, and the fi positive and on an order of magnitude smaller than the di in absolute value. As a result, the ei are negative. Rearranging to put the Jacobian matrix's main diagonal terms nonzero,

$$dD_i - (ei * P^{ei-1} * D_{i-}^{ci})dP = (ci * P^{ei} * D_{i-}^{ci-1})dD_{i-}, i = 1, 2, \ldots, 7. \quad (4.2.1)\text{–}(4.2.7)$$

Regional Non-OPEC Supply Equations

Each of the non-OPEC supply regions has a supply relationship of the form

$$S_j = P^{bj} * S_{j-}^{aj},$$

where S_j = supply in region j-7, S_{j-} = supply in region j-7 lagged one period, bj = the price elasticity of supply in j-7, and aj = a speed of dynamic adjustment factor in j-7. The price elasticities of supply, bj, are all positive. Indices are set to put main diagonal terms nonzero (i.e., in SF1):

$$dS_j - \left(bj * P^{bj-1} * S_{j-}^{aj}\right)dP = \left(aj * P^{bj} * S_{j-}^{aj-1}\right)dS_{j-}, j = 8, 10, \ldots, 15. \quad (4.2.8)\text{–}(4.2.15)$$

Demand for OPEC Oil

The demand for OPEC oil (DO) is the sum of all regional demands net of regional supplies,

$$\text{DO} = \Sigma_i D_i - \Sigma_j S_j.$$

The equation is expressed to keep the diagonal terms nonzero:

$$d\mathrm{DO} - \Sigma_i dD_i + \Sigma_j dS_j = 0. \tag{4.2.16}$$

OPEC Capacity Utilization

OPEC's capacity utilization (CAPUT) is OPEC demand's proportion of (exogenous) maximum OPEC production (MAXCAP).

$$\mathrm{CAPUT} = \mathrm{DO}/\mathrm{MAXCAP}.$$

The equation is expressed to keep the diagonal terms nonzero:

$$d\mathrm{CAPUT} - d\mathrm{DO}/\mathrm{MAXCAP} = -(\mathrm{DO}/\mathrm{MAXCAP}^2)d\mathrm{MAXCAP}. \tag{4.2.17}$$

OPEC Price Evolution Relationship

$$Z = \alpha + \beta/(1 - \mathrm{CAPUT}),$$

where $Z =$ the percentage change (in decimal) of this year's world oil price compared to last year's, and $\{\alpha, \beta\}$ are constant and the slope parameters of the OPEC price reaction function. The equation is expressed to keep the diagonal terms nonzero:

$$dZ + \left(\beta/(1 - \mathrm{CAPUT})^2\right)d\mathrm{CAPUT} = 0. \tag{4.2.18}$$

World Oil Price Determination

This year's price (P) is last year's price (P_-), growing at the rate Z,

$$P = P_-(1 + Z).$$

The equation is expressed to keep the diagonal terms nonzero:

$$dP - P_- dZ = (1 + Z)dP_-. \tag{4.2.19}$$

Summary of Model

The system of equations is summarized below:

Equation(s)	Variable with Same Number	Parameters
(4.2.1)–(4.2.7)	$dD_i : i = 1, 2, \ldots, 7$	$dD_{i-} : i = 1, 2, \ldots, 7$
(4.2.8)–(4.2.15)	$dS_j; j = 8, 9, \ldots, 15$	$dS_{j-} : j = 8, 9, \ldots, 15$
(4.2.16)	$d\mathrm{DO}$	None
(4.2.17)	$d\mathrm{CAPUT}$	$d\mathrm{MAXCAP}$
(4.2.18)	dZ	None.
(4.2.19)	dP	dP_-

The Signed Jacobian Matrix

The signs of the entries of the Jacobian matrix are unambiguous and are found by applying the signs of the coefficient values to the total differentiation of the OMS equations as given above. The results, given as sgn A, are shown in Table 4.1. Inspection of the array reveals that it is sparse, but irreducible. The inference structure for OMS can be readily inspected in the corresponding signed digraph. For the purposes of a qualitative analysis, the array given in Table 4.1 must be further manipulated such that the (nonzero and all positive) main diagonal terms are negative. For the case here, this is accomplished by taking the negative of all entries in the array (i.e., multiplying every column by -1). Given this, the corresponding signed digraph is shown in Figure 4.1.

The nonzeros, a_{ij}, in the array can be interpreted to communicate the circumstance of directed inference: "variable *i depends on* variable *j*." In the corresponding signed digraph, this circumstance is represented by (vertex j) $\rightarrow$ $\pm$(vertex i), with the sign of the directed arc equal to the sign of the corresponding nonzero coefficient. The irreducible nature of the array, i.e., the fact that all of the variables must be solved

TABLE 4.1
Sgn A for OMS

ROW/ COL	1	2	3	4	5	6	7	8	9	10	11	12	13	14	15	16	17	18	19
1	1	0	0	0	0	0	0	0	0	0	0	0	0	0	0	0	0	0	1
2	0	1	0	0	0	0	0	0	0	0	0	0	0	0	0	0	0	0	1
3	0	0	1	0	0	0	0	0	0	0	0	0	0	0	0	0	0	0	1
4	0	0	0	1	0	0	0	0	0	0	0	0	0	0	0	0	0	0	1
5	0	0	0	0	1	0	0	0	0	0	0	0	0	0	0	0	0	0	1
6	0	0	0	0	0	1	0	0	0	0	0	0	0	0	0	0	0	0	1
7	0	0	0	0	0	0	1	0	0	0	0	0	0	0	0	0	0	0	1
8	0	0	0	0	0	0	0	1	0	0	0	0	0	0	0	0	0	0	−1
9	0	0	0	0	0	0	0	0	1	0	0	0	0	0	0	0	0	0	−1
10	0	0	0	0	0	0	0	0	0	1	0	0	0	0	0	0	0	0	−1
11	0	0	0	0	0	0	0	0	0	0	1	0	0	0	0	0	0	0	−1
12	0	0	0	0	0	0	0	0	0	0	0	1	0	0	0	0	0	0	−1
13	0	0	0	0	0	0	0	0	0	0	0	0	1	0	0	0	0	0	−1
14	0	0	0	0	0	0	0	0	0	0	0	0	0	1	0	0	0	0	−1
15	0	0	0	0	0	0	0	0	0	0	0	0	0	0	1	0	0	0	−1
16	−1	−1	−1	−1	−1	−1	−1	1	1	1	1	1	1	1	1	1	0	0	0
17	0	0	0	0	0	0	0	0	0	0	0	0	0	0	0	−1	1	0	0
18	0	0	0	0	0	0	0	0	0	0	0	0	0	0	0	0	−1	1	0
19	0	0	0	0	0	0	0	0	0	0	0	0	0	0	0	0	0	−1	1

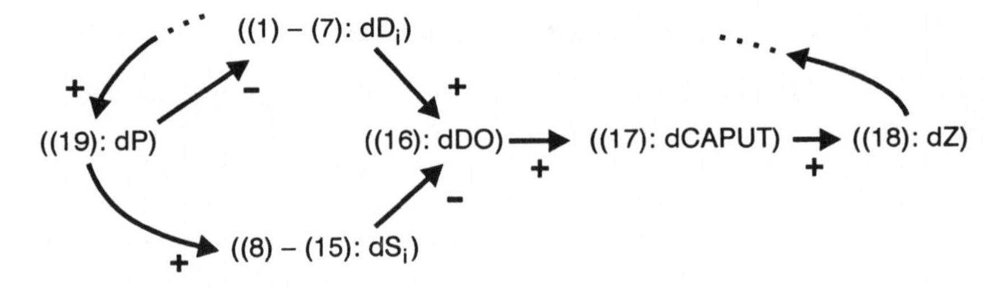

Figure 4.1. SDG(A) corresponding to sgn A in standard form for OMS.

for simultaneously, is revealed by the circumstance that any vertex in the digraph can be reached from any other vertex by following a sequence of directed arcs (i.e., a path in the digraph). Since sgn A is irreducible, all entries of sgn A^{-1} can be nonzero. A cycle in the digraph is a path from a vertex to itself in which any other vertex appears no more than once. The signs of cycles and paths are equal to the signs of the products of the nonzero coefficients that specify the cycles and paths.

The reader can confirm that there are fifteen cycles, all with negative signs; e.g., there are seven cycles of the form

$$(19) \rightarrow_{-} (\text{one of } (1) - (7)) \rightarrow_{+} (16) \rightarrow_{+} (17) \rightarrow_{+} (18) \rightarrow_{+} (19).$$

As a result, the Jacobian matrix is qualitatively invertible. Further, all of the paths from (19) to (16) have the same sign, and there is only one path between any other pair of vertices. As a result, all paths between pairs of vertices have the same sign and all entries of sgn A^{-1} can be signed (and are equal to the negative of the sign of the corresponding path [Maybee and Quirk (1969)]. These results require sgn A to be in standard form. To get the actual signs in sgn A^{-1}, the inverse must be taken back to its form prior to standardization. Here, this is accomplished by multiplying each row of the standardized inverse by -1 (the opposite of the column sign changes needed to set the Jacobian's main diagonal terms all negative). Alternatively, the signed adjoint could be found directly by repeating the analysis for each cofactor (which clearly would be facilitated by computer). Since the main diagonal of the untransformed Jacobian is all positive, the adjoint and inverse matrices have the same sign pattern, as shown in Table 4.2. The correspondence between columns (i.e., equations) and parameter differentials in the

TABLE 4.2
Sgn Adjoint(A) = Sgn A^{-1} for OMS

ROW/ COL	1	2	3	4	5	6	7	8	9	10	11	12	13	14	15	16	17	18	19
1	1	−1	−1	−1	−1	−1	−1	1	1	1	1	1	1	1	1	−1	−1	−1	−1
2	−1	1	−1	−1	−1	−1	−1	1	1	1	1	1	1	1	1	−1	−1	−1	−1
3	−1	−1	1	−1	−1	−1	−1	1	1	1	1	1	1	1	1	−1	−1	−1	−1
4	−1	−1	−1	1	−1	−1	−1	1	1	1	1	1	1	1	1	−1	−1	−1	−1
5	−1	−1	−1	−1	1	−1	−1	1	1	1	1	1	1	1	1	−1	−1	−1	−1
6	−1	−1	−1	−1	−1	1	−1	1	1	1	1	1	1	1	1	−1	−1	−1	−1
7	−1	−1	−1	−1	−1	−1	1	1	1	1	1	1	1	1	1	−1	−1	−1	−1
8	1	1	1	1	1	1	1	1	−1	−1	−1	−1	−1	−1	−1	1	1	1	1
9	1	1	1	1	1	1	1	−1	1	−1	−1	−1	−1	−1	−1	1	1	1	1
10	1	1	1	1	1	1	1	−1	−1	1	−1	−1	−1	−1	−1	1	1	1	1
11	1	1	1	1	1	1	1	−1	−1	−1	1	−1	−1	−1	−1	1	1	1	1
12	1	1	1	1	1	1	1	−1	−1	−1	−1	1	−1	−1	−1	1	1	1	1
13	1	1	1	1	1	1	1	−1	−1	−1	−1	−1	1	−1	−1	1	1	1	1
14	1	1	1	1	1	1	1	−1	−1	−1	−1	−1	−1	1	−1	1	1	1	1
15	1	1	1	1	1	1	1	−1	−1	−1	−1	−1	−1	−1	1	1	1	1	1
16	1	1	1	1	1	1	1	−1	−1	−1	−1	−1	−1	−1	−1	1	−1	−1	−1
17	1	1	1	1	1	1	1	−1	−1	−1	−1	−1	−1	−1	−1	1	1	−1	−1
18	1	1	1	1	1	1	1	−1	−1	−1	−1	−1	−1	−1	−1	1	1	1	−1
19	1	1	1	1	1	1	1	−1	−1	−1	−1	−1	−1	−1	−1	1	1	1	1

array is as follows:

Column(s)	Parameter
(1)–(7)	Lagged demand, dD_{i-}, $i = 1, 2, \ldots, 7$
(8)–(15)	Lagged supply, dS_{j-}, $j = 8, 9, \ldots, 15$
(16)	No parameter in equation (16)
(17)	The negative of maximum OPEC capacity, $-d\text{MAXCAP}$
(18)	No parameter in equation (18)
(19)	The lagged world oil price, dP_{-}

The entry sgn a_{ij}^{-1} gives the direction of influence of the parameter corresponding to the jth column upon the variable corresponding to the ith row. For example, the upper left 7×7 block of sgn A^{-1} is positive for all on-diagonal entries and negative for all off-diagonal entries. This means that for any solution to the OMS, regional demand varies directly with the region's own lagged demand and inversely with each other region's lagged demand (independent of the magnitudes of the coefficients). As another example, inspection of the last column of sgn A^{-1} reveals that when the lagged world oil price is larger, the current solution(s) for regional demands are all smaller, regional supplies are all larger, net OPEC demand is smaller, current OPEC capacity utilization is smaller, the growth rate for the world oil price is smaller, and the solution value for the current world oil price is larger. Each parameter appears in only one equation. As a result, the signs of the variable parameter sensitivities are given by the entries of the signed Jacobian inverse. If a parameter appears in more than one equation, the sign of a variable/parameter sensitivity may not be determinable because the corresponding row in [sgn A^{-1} sgn b] may have both a positive and a negative entry for a single entry of sgn b nonzero. Lady (1983) showed that for sgn A irreducible, *all* of the entries of sgn x could not be qualitatively determined if more than one entry of sgn b was nonzero (see the discussion of Klein's Model I below).

4.4 A QUALITATIVE ANALYSIS OF THE OIL AND GAS SUPPLY MODEL

The Oil and Gas Supply Model (OGSM)[4] is a module in the U.S. Department of Energy's National Energy Modeling System (NEMS). The purpose of the model is to provide price-sensitive projections of domestic oil and gas supply for integration with other estimates of energy supply and demand within NEMS. The OGSM is designed to project supply from currently developed oil and gas resources and to project the economic propensity for and pace of resource development on the basis of current and expected energy prices and estimates of available but undeveloped oil and gas resources.

The OSGM is specified in terms of alternative regions (six regions comprising the lower 48 states, three offshore regions, and three Alaskan regions), fuel types (oil, shallow gas, deep gas, tight sands gas, Devonian shale gas, coal bed methane), and well types (exploratory and developmental). In addition, the model accounts for aspects of foreign trade in natural gas and develops projections through the year 2010. This is a lot of detail. A full-solution iteration of the model for all forecast years would involve many thousands of values for the model's variables and parameters. Generally, from the standpoint of model structure, regional and fuel-type specific solutions are independent of each other. Accordingly, for the purposes of this example, the model was aggregated to account for one region, "the lower 48 United States," one fuel type, "oil," and one time period. Given this, the aggregated form of OGSM presented here has thirty-four variables and thirty-nine parameters.

Getting the System into Standard Form

The OGSM is presented as a system of thirty-four equations of the form

$$\text{variable}_i = f^i(\text{variable(s)}, \text{parameter(s)}), i = 1, 2, \ldots, 34.$$

The presentation of the model in this form facilitates the qualitative analysis in that variables and equations can be assigned indices such that each variable can appear as the left variable in an equation with the same index. For constructing the signed Jacobian, the equations are rearranged into the form

$$-\text{variable}_i i + f^i(\text{variable(s)}, \text{parameter(s)}) = 0, i = 1, 2, \ldots, 34.$$

Given this, the main diagonal of the model's Jacobian matrix is all negative, and the analysis can proceed directly without further manipulation in order to put the Jacobian matrix into the required standard form.

Format of Specification

The mathematical description of the model given below is based upon appendix B of OGSM (1994). In what follows, the format is to list the variable corresponding to each equation in the left-hand column. The variables and parameters on which the variable depends are listed in the right-hand column. Each variable has the same index as the equation in which it appears. When a variable appears as an explanatory variable in an equation other than the equation to which it corresponds (i.e., has the same index), its index is prefixed with a "V." The magnitudes identified in the right-hand column with an index not prefixed with a "V" are parameters. For each explanatory variable or parameter that appears in the equation, the sign of its derivative is given in parentheses. These signs are used directly in forming the signed Jacobian matrix. For example, equation 2 on page B-3 of OGSM (1994) takes the aggregated form

$$\text{PROJDCFON(I = 2)} = \text{SUCDCFON(I = 2)} + [(1/SR) - 1]^{*}\text{DRYDCFON(I = 2)}.$$

This equation is organized in the model specification below as follows:

Discounted Cash Flow for Representative Onshore Project

4. PROJDCFON(I = 2) V2. SUCDCFON(I = 2, +)
V3. DRYDCFON(I = 2, −)
11. SR(I = 2, +): SUCCESS RATE

This presentation is intended to communicate the following information:

The variable measures the discounted cash flow for a representative onshore project.

The variable name is PROJDCFON(I = 2), which is variable 4 in the listing given here. The notation (I = 2) indicates a development well; in various other listings, (I = 1) indicates an exploratory well, and (I = 1, s) or (I = 2, s) also indicates the year of the project.

SUCDCFON(I = 2) is an explanatory variable in this equation, with a positive derivative that appears as variable 2 in the listing given here.

DRYDCFON(I = 2) is an explanatory variable in this equation, with a negative derivative that appears as variable 3 in the listing given here.

SR(I = 2) is an explanatory variable in this equation, with a positive derivative that is parameter 11 in the listing given here. Since SR(I = 2) has not appeared in an earlier equation in the listing here, it is annotated as "SUCCESS RATE" (each variable and parameter is briefly described the first time it appears).

The thirty-four equations of the aggregated OGSM are enumerated below in this format. The corresponding signed Jacobian matrix has an all-negative main diagonal with nonzero off-diagonal terms of the signs given in the right-hand column for magnitudes with indices prefixed by a "V."

Lower 48 Oil and Gas Supply Model Specification (Onshore)

Net Cash Flow for Onshore Project For a Given Year

1. NCFON(I = 2, s)
    1. REV(I = 2, s, +): REVENUE FROM SALES
    2. ROY(I = 2, s, −): ROYALTY TAXES
    3. PRODTAX(I = 2, s, −): PRODUCTION TAXES (SEV., AD.VAL.)
    4. DRILLCOST(I = 2, s, −): DRILLING COSTS
    5. EQUIPCOST(I = 2, s, −): LEASE EQUIP. COST
    6. OPCOST(I = 2, s, −): OPERATING COSTS
    7. DRYCOST(I = 2, s, −): COST OF DRY WELL
    8. STATETAX(I = 2, s, −): STATE INCOME TAX LIABILITY
    9. FEDTAX(I = 2, s, −): FEDERAL INCOME TAX LIABILITY

Discounted Cash Flow for a Successful Well

2. SUCDCFON(I = 2)
    - V1. NCFON(I = 2, s, +)
    - 10. DR(−): DISCOUNT RATE (IMPLIED)

Discounted Cash Flow for a Dry Hole

3. DRYDCFON(I = 2)
    - V1. NCFON(I = 2, s, +)
    - 10. DR(+)

Discounted Cash Flow for Representative Onshore Project

4. PROJDCFON(I = 2)
    - V2. SUCDCFON(I = 2, +)
    - V3. DRYDCFON(I = 2, −)
    - 11. SR(I = 2, +): SUCCESS RATE

Discounted Cash Flow for Onshore Development Project

5. DCFON(I = 2)	V.4 PROJDCFON(I = 2, +) 11. SR(I = 2, +)

Discounted Cash Flow for Exploratory Project

6. PROJDCFON(I = 1)	12. SUCDCFON(I = 1, +): CASH FLOW FOR SUC. EX. WELL[5] 13. DRYDCFON(I = 1, −): CASH FLOW FOR DRY EX. WELL V2. SUCDCFON(I = 2, +) V3. DRYDCFON(I = 2, −) 14. M(+): NUMBER DEV. WELLS IN PROJECT 11. SR(I = 2, +) 15. SR(I = 1, +): SUCCESS RATE FOR EX. WELLS

Discounted Cash Flow for Onshore Exploratory Project

7. DCFON(I = 1)	V6. PROJDCFON(I = 1, +) 15. SR(I = 1, +)

"National" Weighted Averages (which degenerate to the above project-specific totals, given that regional detail has been aggregated)

8. NDCFON(I = 1)	V7. DCFON(I = 1, +)
9. NDCFON(I = 2)	V5. DCFON(I = 2, +)

National Drilling Expenditures

10. NSPENDON(I = 1)	16. a0−a3: PARAMETERS V8. NDCFON(I = 1, +) V12. CF(+): INDUSTRY CASH FLOW/$ FROM OIL &GAS OP.
11. NSPENDON(I = 2)	17. b0−b3: PARAMETERS V9. NDCFON(I = 2, +) 18. NSPENDON(I = 2, LAG, +): LAGGED EXPENDITURE V12. CF(+)

Industry Cash Flow

12. CF	19. e0–e1: PARAMETERS
	20. WOP(+): WORLD OIL PRICE

Number of Wells Drilled

13. WELLSON(I = 1)	V10. NSPENDON(I = 1, +)
	V15. COST(I = 1, −): DRILLING COST/WELL
14. WELLSON(I = 2)	V11. NSPENDON(I = 2, +)
	V16. COST(I = 2, −)

Cost per Well

15. COST(I = 1)	21. TECH(I = 1, −): TECHNOLOGICAL CHANGE FACTOR
	22. COST(I = 1, 1990, +): BASE YEAR (1990) COST PER WELL
	23. t(−): TIME
16. COST(I = 2)	24. TECH(I = 2, −): AS ABOVE
	25. COST(I = 2, 1990, +)
	23. t(−)

Number of Successful Wells

17. SUCWELSON(I = 1)	V13. WELLSON(I = 1, +)
	15. SR(I = 1, +)
18. SUCWELSON(I = 2)	V14. WELLSON(I = 2, +)
	11. SR(I = 2, +)

Number of Dry Wells

19. DRYWELON(I = 1)	V13. WELLSON(I = 1, +)
	V17. SUCWELSON(I = 1, −)
20. DRYWELON(I = 2)	V14. WELLSON(I = 2, +)
	V18. SUCWELSON(I = 2, −)

Number of Exploratory New Field Wildcats

21. SW1 — 26. $\alpha(+)$: PROPORTION OF SUCWELSON(I = 1)
V17. SUCWELSON(I = 1, +)

Number of Other Exploratory Wells

22. SW2 — 26. $\alpha(-)$
V17. SUCWELSON(I = 1, +)

Number of Developmental Wells

23. SW3 — V18. SUCWELSON(I = 2, +):
SW3 = SUCWELSON(I = 2)

New Field Wildcats Finding Rate

24. FR1 — 27. FR1(LAG, +): FINDING RATE FOR $t-1$
V25. d1(−): FINDING RATE DECLINE PARAMETER
V21. SW1(−): SUCCESSFUL NEW FIELD WILDCATS

Finding Rate Decline Parameter

25. d1 — 27. FR1(LAG, +)
31. FRMIN(−): MINIMUM ECON. FINDING RATE
28. Q(−): INITIAL RECOVERABLE RESOURCE EST.
29. TECH(−): TECHNOLOGY FACTOR
23. t(−)
30. CUMRES(+): CUMULATIVE RESERVE DISCOVERIES

Change in Proved Reserves

26. $_{\Delta}$R — 32. X(−): RESERVES GROWTH FACTOR
V21. SW1(+)

Other Exploratory Wells Finding Rate

27. FR2 — 33. FR2(LAG, +): FINDING RATE FOR $t-1$
V29. d2(−): OTHER EXP. WELLS DECLINE PARAMETER
V22. SW2(−): SUCCESSFUL OTHER EXP. WELLS

Developmental Wells Finding Rate

28. FR3
   - 34. FR3(LAG, +)
   - V30. d3(−): DEVELOPMENTAL WELL DECLINE PARAMETER
   - V23. SW3(−): SUCCESSFUL DEVELOPMENTAL WELLS

Other Exploratory Wells Decline Parameter

29. d2
   - 33. FR2(LAG, +)
   - 35. I(−): INITIAL INFERRED RESERVES
   - 29. TECH(−)
   - 23. t(−)
   - 32. X(−)
   - 27. FR1(LAG, −)
   - V21. SW1(−)
   - 33. FR2(LAG, +)
   - V22. SW2(+)
   - 34. FR3(LAG, +)
   - V23. SW3(+)
   - 37. DECFAC(+): DECLINE RATE ADJUSTMENT FACTOR

Developmental Wells Decline Parameter

30. d3
   - 34. FR3(LAG, +)
   - 35. I(−): INITIAL INFERRED RERSERVES
   - 29. TECH(−)
   - 23. t(−)
   - 32. X(−)
   - 27. FR1(LAG, −)
   - V21. SW1(−)
   - 33. FR2(LAG, +)
   - V22. SW2(+)
   - 34. FR3(LAG, +)
   - V23. SW3(+)
   - 37. DECFAC(+): DECLINE RATE ADJUSTMENT FACTOR

Reserve Additions

31. RA	32. X(−)
	V24. FR1(+)
	V21. SW1(+)
	V27. FR2(+)
	V22. SW2(+)
	V28. FR3(+)
	V23. SW3(+)

End-of-Year Proved Reserves

32. R	36. R(LAG, +): START OF YEAR RESERVES
	39. Q*(−): PRODUCTION[6]
	V31. RA(+)

Production / Reserves Ratio

33. PR	39. Q*(+)
	36. R(LAG, −)

Expected Production / Reserves Ratio

34. PR(LEAD)	36. R(LAG, +)
	V33. PR(+)[7]
	38. PRNEW(+)
	V31. RA(+)
	V32. R(−)

The Signed Jacobian Matrix

Inspection of these equations reveals that in terms of the documentation provided, six variables are independent of one another and any of the other variables. These are variables 1, 12, 15, 16, 25, and 33. The Jacobian matrix and its inverse were reordered compared to the above order of presentation such that the first six rows and columns correspond to these fully independent variables. The remaining twenty-eight variables correspond one-to-one to the remaining twenty-eight rows and columns. The signed Jacobian matrix is shown in Table 4.3. (entries not shown are all equal to zero).

For interpretation, consider variable 4, PROJDCFON(I = 2), used as an example above. As a result of the reordering of equations to isolate the six fully independent variables, this variable and its equation correspond to the ninth row of the Jacobian. Since only variables 2 and 3 appear in this equation in addition to variable 4 itself, all of the remaining entries in this row of the Jacobian are equal to zero. From the convention above, the diagonal entry (here, entry [9, 9]) is always negative. The remaining nonzeros are given the signs of their derivatives, as indicated in the equation corresponding to variable #4; thus

TABLE 4.3
Jacobian Matrix for OGSM

ROW/ COL	1	12	15	16	25	33	2	3	4	5	6	7	8	9	10	11	13	14	17	18
1	−1	0	0	0	0	0	0	0	0	0	0	0	0	0	0	0	0	0	0	0
12	0	−1	0	0	0	0	0	0	0	0	0	0	0	0	0	0	0	0	0	0
15	0	0	−1	0	0	0	0	0	0	0	0	0	0	0	0	0	0	0	0	0
16	0	0	0	−1	0	0	0	0	0	0	0	0	0	0	0	0	0	0	0	0
25	0	0	0	0	−1	0	0	0	0	0	0	0	0	0	0	0	0	0	0	0
33	0	0	0	0	0	−1	0	0	0	0	0	0	0	0	0	0	0	0	0	0
2	1	0	0	0	0	0	−1	0	0	0	0	0	0	0	0	0	0	0	0	0
3	1	0	0	0	0	0	0	−1	0	0	0	0	0	0	0	0	0	0	0	0
4	0	0	0	0	0	0	1	−1	−1	0	0	0	0	0	0	0	0	0	0	0
5	0	0	0	0	0	0	0	0	1	−1	0	0	0	0	0	0	0	0	0	0
6	0	0	0	0	0	0	1	−1	0	0	−1	0	0	0	0	0	0	0	0	0
7	0	0	0	0	0	0	0	0	0	0	1	−1	0	0	0	0	0	0	0	0
8	0	0	0	0	0	0	0	0	0	0	0	1	−1	0	0	0	0	0	0	0
9	0	0	0	0	0	0	0	0	0	1	0	0	0	−1	0	0	0	0	0	0
10	0	1	0	0	0	0	0	0	0	0	0	0	1	0	−1	0	0	0	0	0
11	0	1	0	0	0	0	0	0	0	0	0	0	0	1	0	−1	0	0	0	0
13	0	0	−1	0	0	0	0	0	0	0	0	0	0	0	1	0	−1	0	0	0
14	0	0	0	−1	0	0	0	0	0	0	0	0	0	0	0	1	0	−1	0	0
17	0	0	0	0	0	0	0	0	0	0	0	0	0	0	0	0	1	0	−1	0
18	0	0	0	0	0	0	0	0	0	0	0	0	0	0	0	0	0	1	0	−1
19	0	0	0	0	0	0	0	0	0	0	0	0	0	0	0	0	1	0	−1	0
20	0	0	0	0	0	0	0	0	0	0	0	0	0	0	0	0	0	1	0	−1
21	0	0	0	0	0	0	0	0	0	0	0	0	0	0	0	0	0	0	1	0
22	0	0	0	0	0	0	0	0	0	0	0	0	0	0	0	0	0	0	1	0
23	0	0	0	0	0	0	0	0	0	0	0	0	0	0	0	0	0	0	0	1
24	0	0	0	0	−1	0	0	0	0	0	0	0	0	0	0	0	0	0	0	0
26	0	0	0	0	0	0	0	0	0	0	0	0	0	0	0	0	0	0	0	0
27	0	0	0	0	0	0	0	0	0	0	0	0	0	0	0	0	0	0	0	0
28	0	0	0	0	0	0	0	0	0	0	0	0	0	0	0	0	0	0	0	0
29	0	0	0	0	0	0	0	0	0	0	0	0	0	0	0	0	0	0	0	0
30	0	0	0	0	0	0	0	0	0	0	0	0	0	0	0	0	0	0	0	0
31	0	0	0	0	0	0	0	0	0	0	0	0	0	0	0	0	0	0	0	0
32	0	0	0	0	0	0	0	0	0	0	0	0	0	0	0	0	0	0	0	0
34	0	0	0	0	0	1	0	0	0	0	0	0	0	0	0	0	0	0	0	0

TABLE 4.3—(*Continued*)

ROW/ COL	19	20	21	22	23	24	26	27	28	29	30	31	32	34
19	−1	0	0	0	0	0	0	0	0	0	0	0	0	0
20	0	−1	0	0	0	0	0	0	0	0	0	0	0	0
21	0	0	−1	0	0	0	0	0	0	0	0	0	0	0
22	0	0	0	−1	0	0	0	0	0	0	0	0	0	0
23	0	0	0	0	−1	0	0	0	0	0	0	0	0	0
24	0	0	−1	0	0	−1	0	0	0	0	0	0	0	0
26	0	0	1	0	0	0	−1	0	0	0	0	0	0	0
27	0	0	0	−1	0	0	0	−1	0	−1	0	0	0	0
28	0	0	0	0	−1	0	0	0	−1	0	−1	0	0	0
29	0	0	−1	1	1	0	0	0	0	−1	0	0	0	0
30	0	0	−1	1	1	0	0	0	0	0	−1	0	0	0
31	0	0	1	1	1	1	0	1	1	0	0	−1	0	0
32	0	0	0	0	0	0	0	0	0	0	0	1	−1	0
34	0	0	0	0	0	0	0	0	0	0	0	1	−1	−1

the derivative with respect to variable 2 is positive (entry [7,9] = 1), and the derivative with respect to variable 3 is negative (entry [8,9] = −1).

The Inverse Jacobian Matrix

The OMS is a fully interdependent system. As a result, all of the entries in the corresponding inverse Jacobian matrix are, or at least could be, nonzero. Although the (aggregated) OGSM presented here has almost twice as many variables, its dependence structure is such that no group of variables is simultaneously interdependent. As a result, the variables can be solved for one at a time. Given this, the corresponding Jacobian matrix is strictly triangular. That is, the equations and variables can be indexed such that all entries above the main diagonal are zero entries. These entries might be called "logical zeros." Inspection of the signed Jacobian shown in Tables 4.3 reveals that the given arrangement conforms to this pattern except for the rows/columns indexed 27 and 28. These rows/columns are just slightly out of order, and the strictly triangular pattern would be achieved if they were swapped with, respectively, the rows/columns indexed 24 and 26. The entries of the inverse Jacobian matrix corresponding to the logical zeros of the Jacobian matrix must also be equal to zero, independent of the values of the nonzero entries of the Jacobian matrix.

The interdependence among the variables and parameters is revealed through inspection of the inverse Jacobian matrix. In principle, this 34×34 array has 1,156 entries. Since the OGSM as presented here is strictly triangular, 561 of these entries are zero because of the model's structure. Of the remaining 595, 232 are also zero because of the further independence of variables within the triangular structure; i.e., each variable in the above solution roster does not depend upon *all* of the variables that come earlier in the listing. These zeros might also be called "logical zeros." Of the remaining 363 nonzeros, the signs of 212 are determinable from the signs in the Jacobian matrix, independent of the magnitudes of the model's coefficients. The signs of the remaining 151 cannot be determined from the signed Jacobian alone.

The qualitatively indeterminate entries might be resolvable through the addition of additional conditions for the entries of the Jacobian matrix, so-called side conditions. The use of side conditions is explicitly discussed in the Section 4.5. Looking ahead to this class of issue, for the qualitatively indeterminate entries of OGSM's inverse Jacobian matrix, one might wish to assess the severity of indeterminacy or the degree to which side conditions must be formulated in order to resolve the indeterminacy of the inverse entry's sign. Given that the Jacobian matrix is qualitatively invertible (because of its triangular structure), the residual issue for signing an entry in the Jacobian inverse can be investigated by constructing the array for the corresponding cofactor, manipulating it into standard form, and then assessing the signs of cycles in the corresponding signed digraph. If all cycles are negative, then the entry can be signed. If any cycles are positive, then the entry is qualitatively indeterminate. For the indeterminate entries, a first-order indicator of the degree of indeterminacy could be the number of positive cycles found. In general, the more positive cycles the worse.

This measure is refined in the Section 4.5, but is used here for the indeterminate entries in the OGSM inverse. The convention used is as follows:

Entries assigned a 1, 0, or -1 are qualitatively determinate with signs that are $+$, 0, or $-$, respectively.

All other entries are qualitatively indeterminate. They are assigned a value equal to the number of positive cycles minus one found in the array corresponding to the entry's cofactor (e.g., a value of 3 represents an array with two positive cycles).

Using this convention, the inverse Jacobian matrix for OGSM is shown in Table 4.4. The manner of interpreting the signed inverse Jacobian is as follows. First, note that the sign convention of the Jacobian set the differential of each variable negative in the equation corresponding to the variable. As a result, the intuitive direction of influence in the inverse Jacobian is the negative of the entry given. As an example, consider the second column in the first two arrays in Table 4.4. This column corresponds to variable 12, petroleum industry cash flow. In this equation, the industry cash flow is estimated as an exponential (i.e., log linear) function of the world oil price (WOP),

$$CF = e0 * \mathrm{WOP}^{e1},$$

Thus, the entries in the second column of the Jacobian inverse show the (negative of the) direction of influence on each of the thirty-four variables of changes in the WOP. Inspection reveals the following:

1. Variables 1, 15, 16, 25, 33, and 2–9 are independent of the WOP, and the remaining variables are dependent upon the WOP.

TABLE 4.4
Inverse Jacobian Matrix for OGSM

ROW/ COL	1	12	15	16	25	33	2	3	4	5	6	7	8	9	10	11	13	14	17	18
1	−1	0	0	0	0	0	0	0	0	0	0	0	0	0	0	0	0	0	0	0
12	0	−1	0	0	0	0	0	0	0	0	0	0	0	0	0	0	0	0	0	0
15	0	0	−1	0	0	0	0	0	0	0	0	0	0	0	0	0	0	0	0	0
16	0	0	0	−1	0	0	0	0	0	0	0	0	0	0	0	0	0	0	0	0
25	0	0	0	0	−1	0	0	0	0	0	0	0	0	0	0	0	0	0	0	0
33	0	0	0	0	0	−1	0	0	0	0	0	0	0	0	0	0	0	0	0	0
2	−1	0	0	0	0	0	−1	0	0	0	0	0	0	0	0	0	0	0	0	0
3	−1	0	0	0	0	0	0	−1	0	0	0	0	0	0	0	0	0	0	0	0
4	2	0	0	0	0	0	−1	1	−1	0	0	0	0	0	0	0	0	0	0	0
5	2	0	0	0	0	0	−1	1	−1	−1	0	0	0	0	0	0	0	0	0	0
6	2	0	0	0	0	0	−1	1	0	0	−1	0	0	0	0	0	0	0	0	0
7	2	0	0	0	0	0	−1	1	0	0	−1	−1	0	0	0	0	0	0	0	0
8	2	0	0	0	0	0	−1	1	0	0	−1	−1	−1	0	0	0	0	0	0	0
9	2	0	0	0	0	0	−1	1	−1	−1	0	0	0	−1	0	0	0	0	0	0
10	2	−1	0	0	0	0	−1	1	0	0	−1	−1	−1	0	−1	0	0	0	0	0
11	2	−1	0	0	0	0	−1	1	−1	−1	0	0	0	−1	0	−1	0	0	0	0
13	2	−1	1	0	0	0	−1	1	0	0	−1	−1	−1	0	−1	0	−1	0	0	0
14	2	−1	0	1	0	0	−1	1	−1	−1	0	0	0	−1	0	−1	0	−1	0	0
17	2	−1	1	0	0	0	−1	1	0	0	−1	−1	−1	0	−1	0	−1	0	−1	0
18	2	−1	0	1	0	0	−1	1	−1	−1	0	0	0	−1	0	−1	0	−1	0	−1
19	3	2	2	0	0	0	2	2	0	0	2	2	2	0	2	0	2	0	1	0
20	3	2	0	2	0	0	2	2	2	2	0	0	0	2	0	2	0	2	0	1
21	2	−1	1	0	0	0	−1	1	0	0	−1	−1	−1	0	−1	0	−1	0	−1	0
22	2	−1	1	0	0	0	−1	1	0	0	−1	−1	−1	0	−1	0	−1	0	−1	0
23	2	−1	0	1	0	0	−1	1	−1	−1	0	0	0	−1	0	−1	0	−1	0	−1
24	2	1	−1	0	1	0	1	−1	0	0	1	1	1	0	1	0	1	0	1	0
26	2	−1	1	0	0	0	−1	1	0	0	−1	−1	−1	0	−1	0	−1	0	−1	0
27	5	4	3	−1	0	0	2	2	1	1	3	3	3	1	3	1	3	1	3	1
28	4	4	2	−1	0	0	2	2	1	1	2	2	2	1	2	1	2	1	2	1
29	4	3	2	1	0	0	2	2	−1	−1	2	2	2	−1	2	−1	2	−1	2	−1
30	4	3	2	1	0	0	2	2	−1	−1	2	2	2	−1	2	−1	2	−1	2	−1
31	8	6	5	2	1	0	6	6	2	2	5	5	5	2	5	2	5	2	5	2
32	8	6	5	2	1	0	6	6	2	2	5	5	5	2	5	2	5	2	5	2
34	8	7	6	3	2	−1	7	7	3	3	6	6	6	3	6	3	6	3	6	3

TABLE 4.4—(*Continued*)

ROW/ COL	19	20	21	22	23	24	26	27	28	29	30	31	32	34
19	−1	0	0	0	0	0	0	0	0	0	0	0	0	0
20	0	−1	0	0	0	0	0	0	0	0	0	0	0	0
21	0	0	−1	0	0	0	0	0	0	0	0	0	0	0
22	0	0	0	−1	0	0	0	0	0	0	0	0	0	0
23	0	0	0	0	−1	0	0	0	0	0	0	0	0	0
24	0	0	1	0	0	−1	0	0	0	0	0	0	0	0
26	0	0	−1	0	0	0	−1	0	0	0	0	0	0	0
27	0	0	−1	1	1	0	0	−1	0	1	0	0	0	0
28	0	0	−1	1	1	0	0	0	−1	0	1	0	0	0
29	0	0	1	−1	−1	0	0	0	0	−1	0	0	0	0
30	0	0	1	−1	−1	0	0	0	0	0	−1	0	0	0
31	0	0	4	2	2	−1	0	−1	−1	1	1	−1	0	0
32	0	0	4	2	2	−1	0	−1	−1	1	1	−1	−1	0
34	0	0	5	3	3	2	0	2	2	2	2	2	1	−1

2. If the WOP increases, then the solution values for variables 10, 11, 13, 14, 17, 18, 21–23, and 26 will also increase for any values of the OGSM coefficients of the signs assumed here.
3. If the WOP increases, then the solution value for variable 24, the new field wildcats finding rate, will decrease for any values of the OGSM coefficients of the signs assumed here.
4. Although changes in the WOP will influence variables 27–34 (except for 33), the direction of influence is not independent of the magnitudes of the OGSM coefficients. The values given for the entries are equal to the number of positive cycles less one in the signed digraph corresponding to the array of the entry's cofactor when placed in standard form.

Generally, the signs in the inverse Jacobian matrix specify the (negative of the) necessary direction of influence (i.e., direction of change in its solution value) on the variable identified by the row index of each entry exerted by increases in the parameters found in the equation corresponding to the variable identified by the column index of the entry.

4.5 KLEIN'S MODEL I

A small macroeconomic model used to illustrate multistage least squares estimation is sometimes called Klein's Model I.[8] A specification of this model, in which the equations have been rearranged to show the endogenous variables on the left side and the predetermined magnitudes on the right side and listed in order of their correspondence to the endogenous variables chosen here to be ordered (i.e., the equations and variables have been ordered such that the Jacobian has an all-nonzero main diagonal), is as follows:

$$
\begin{aligned}
Y - C - I &= G - T, \\
C - \alpha_2 W_1 - \alpha_1 \Pi &= \alpha_0 + \alpha_2 W_2 + \alpha_2 \Pi_{-1} + u_1, \\
I - \beta_1 \Pi &= \beta_0 + \beta_2 \Pi_{-1} + \beta_3 K_{-1} + u_2, \\
-\gamma_1 Y + W_1 &= \gamma_0 + \gamma_1 (T - W_2) + \gamma_2 (Y + T - W_2)_{-1} + \gamma_3 t + u_3, \\
Y - W_1 - \Pi &= W_2, \\
-I + K &= K_{-1},
\end{aligned}
\tag{4.3}
$$

where the endogenous variables are $Y =$ national income, $C =$ consumption, $I =$ investment, $W_1 =$ private wage bill, $\Pi =$ profits, and $K =$ capital stock; other given magnitudes are $G =$ government purchases, $W_2 =$ government wage bill, $T =$ indirect taxes, and $t =$ time in years; and, u_1, u_2, and u_3 are the error terms associated with the behavioral equations. The Jacobian matrix for this model is

$$
J = [a_{ij}] = \begin{bmatrix}
1 & -1 & -1 & 0 & 0 & 0 \\
0 & 1 & 0 & -\alpha_2 & -\alpha_1 & 0 \\
0 & 0 & 1 & 0 & -\beta_1 & 0 \\
-\gamma_1 & 0 & 0 & 1 & 0 & 0 \\
1 & 0 & 0 & -1 & -1 & 0 \\
0 & 0 & -1 & 0 & 0 & 1
\end{bmatrix}.
$$

Since the main diagonal of J is all nonzero, transformation into standard form requires that all but the fifth column be multiplied by -1. Let J' be J put into SF2 (J' is also in SF3 since the diagonal entries all equal one in absolute value). Inspection of SDG(J') would reveal five distinct cycles, two negative and three positive, each involving the vertex 1 (all cycles are maximal). As a result, J' is not qualitatively invertible; however, since no two cycles are disjoint, J' is *intercyclic* (definition 3.2). Given this, the structure of the matrix will facilitate the use of side conditions.

One set of estimates of the coefficients of the model is $\alpha_1 = .0479$, $\alpha_2 = .817$, $\beta_1 = .2111$, and $\gamma_1 = .4282$. Given this, the values of the positive cycles of SDG(J') are as follows:

$$
\begin{aligned}
v(c_1) &= ((-a_{41} = .4282)(-a_{24} = .817)(-a_{12} = 1)) = .350 \\
v(c_2) &= ((-a_{51} = -1)(a_{25} = -.0479)(-a_{12} = 1)) = .0479 \\
v(c_3) &= ((-a_{51} = -1)(a_{35} = -.2111)(-a_{13} = 1)) = \underline{.2111} \\
& \qquad\qquad\qquad\qquad\qquad\qquad\qquad\qquad\quad .609\ .
\end{aligned}
$$

As a result, from Proposition 3.4, sgn det$(J') = (-1)^n$ (this condition is also satisfied for all of the main diagonal cofactors, which are also in SF3). Inspection of J's cofactors reveals that all of the corresponding arrays are GK-matrices (definition 3.3), and many of these are invertible on the basis of the signs of their entries alone (i.e., they are QI-matrices). In fact, although J' is a relatively small array, the arrays corresponding to the entries of J's adjoint provide examples of many of the various structures discussed in Chapter 3:

$$
\text{Arrays in } J\text{'s Adjoint} = \begin{bmatrix}
\text{QI} & \text{QI} & \text{QI} & \text{KB} & \text{QI} & 0 \\
\text{KA} & \text{KA} & \text{KA} & \text{KB} & \text{QI} & 0 \\
\text{KB} & \text{KB} & \text{GK} & \text{KB} & \text{KB} & 0 \\
\text{QI} & \text{QI} & \text{QI} & \text{KB} & \text{QI} & 0 \\
\text{KB} & \text{KB} & \text{KB} & \text{KB} & \text{KB} & 0 \\
\text{KB} & \text{KB} & \text{GK} & \text{KB} & \text{KB} & \text{GK}
\end{bmatrix}.
$$

The entries are notated as follows (all were transformed into standard form): QI = QI-matrix, KA = K-matrix in form A, KB = K-matrix in form B, and GK = GK-matrix that is not a K-matrix or a QI-matrix.

Each cofactor was analyzed and its sign resolved, with the following side conditions used as required:

(a) The corresponding array is a QI-matrix.
(b) The corresponding array is a K-matrix in form A such that a negative cycle's term is equal to the single positive cycle's term.
(c) $0 < \beta_1 = \partial I/\partial \Pi < 1$.
(d) $0 < \gamma_1 = \partial W_1/\partial Y < 1$.
(e) $0 < \alpha_2 = \partial C/\partial W_1 < 1$.
(f) $0 < \gamma_1 \alpha_2 < 1$.
(g) The corresponding array is a cycle dominant K-matrix.
(h) The corresponding array is a cycle dominant GK-matrix.

Since the matrices are small, the side conditions are expressed directly in terms of the coefficients if practicable, without further putting the cofactor arrays into a standard form.

The algorithm used to put the subarrays into standard form searched for column permutations to fill the main diagonal in the order of columns. Given this, for cases (b), (c), (e), and (f), the single odd-signed term concerns off-diagonal entries and the array is a K-matrix. For these, the main diagonal term is larger than the term corresponding to the single positive cycle. For case (d), the corresponding array is a K-matrix with a single cycle. The term corresponding to the cycle is larger than the main diagonal term.

The cofactor analysis yielded the following array:

$$\operatorname{sgn}(\operatorname{adjoint}(J)) = \begin{bmatrix} -I^a & -I^a & -I^a & -I^g & I^a & 0 \\ -I^b & -I^c & -I^b & -I^g & I^a & 0 \\ -I^d & -I^d & -I^h & I^e & I^f & 0 \\ -I^a & -I^a & -I^a & -I^h & I^a & 0 \\ -I^d & -I^d & -I^d & I^e & I^f & 0 \\ -I^d & -I^d & -I^h & I^e & I^e & -I^h \end{bmatrix},$$

where the superscript letters indicate the side conditions. Since sgn $\det(J) < 0$, the entries of sgn adjoint(J) are the negative of the entries of sgn J^{-1}.

For the model used in this example, the exogenous magnitudes are not each isolated in a distinct equation. As a result, determining the signs of the variable/parameter sensitivities requires a further step to account for interdependencies when parameters appear in more than one equation. For the results here, although the inverse Jacobian matrix has been entirely signed, some of the sensitivities cannot be signed. The resulting determinable signed variable/parameter sensitivities are

$$\operatorname{sgn}\left[\frac{d(\text{variable})}{d(\text{parameter})}\right] = \begin{bmatrix} & T & T_{-1} & W_2 & W_{2-} & t & Y_{-1} & \Pi_{-1} & K_{-1} \\ Y & ? & 1 & ? & -1 & 1 & 1 & 1 & -1 \\ C & ? & 1 & ? & -1 & 1 & 1 & 1 & -1 \\ I & -1 & -1 & ? & 1 & -1 & -1 & 1 & -1 \\ W_1 & ? & 1 & ? & -1 & 1 & 1 & 1 & -1 \\ \Pi & -1 & -1 & ? & 1 & -1 & -1 & 1 & -1 \\ K & -1 & -1 & ? & 1 & -1 & -1 & 1 & ? \end{bmatrix}.$$

The above array provides the basis for auditing the results of sensitivity runs of computer-based versions of the model (which would be more important for a large model). The declining secular trends evidenced in the fifth column might limit the policy analysis applicability of the model for the ranges of values of the estimated coefficients. Of the forty-eight sensitivities, only ten cannot be determined for the ranges of coefficient values assumed. These particular sensitivities are therefore likely to be the most sensitive to the range of error inherent in the estimated model. Since this model is small, there may be other ways to reach similar conclusions about its nature. For larger models, such findings would be extremely difficult. The analysis of the model's dependence structure with side conditions, as presented here, is one way to assess a model's robustness to its constituent forms and magnitudes.

4.6 SUMMARY

The examples presented here represent a wide range of circumstances. The OMS is fully interdependent, and yet the entire inverse Jacobian matrix could be signed on the basis of the Jacobian matrix's sign pattern alone. The OGSM was somewhat larger, but had a completely triangular structure (no two or more variables had to be solved for simultaneously). Even so, some of the entries of the inverse Jacobian matrix could not be found from the Jacobian matrix's sign pattern. Klein's Model I is a small model, and yet side conditions were required to sign its determinant and most of its cofactors. Even so, since parameters appeared in more than one equation, although the inverse Jacobian matrix was entirely signed (given the side conditions), some variable/parameter sensitivities remained ambiguous. Nevertheless, the examples provide an ample cross-section of successful application of nonparametric comparative statics analyses. Actual models, even large ones, tend to have sparse Jacobian matrices, often with repetitive structures. An investigation of the models using qualitative techniques (plus side conditions) can provide substantial insight into how the models operate, test their robustness to ranges of constituent magnitudes, and enable audits of computer-implemented systems compared to their mathematical expression. It is true that the conditions required for a successful analysis appear stringent as a matter of logic, as Samuelson (1947) originally supposed; however, actual systems are often found to conform to the conditions for nonparametric analysis, and the results obtained can become a useful ingredient of understanding and working with models.

5

The Maximization Hypothesis

5.1 INTRODUCTION

Qualitative techniques are useful because theories and models are formulated for a wide range of diverse environments, where quantitative values for parameters are unknown, unreliable, or unstable. Using a qualitatively specified approach permits an investigation into the implications of a theory for a broad class of such environments. In a few exceptional cases, the model does not impose any quantitative or semiquantitative restrictions on parameter values. For those cases, a "purely qualitative" approach, as described in Chapter 2, is applicable. But in most models, there are restrictions imposed on parameter values by the structure of the model, and these restrictions have to be incorporated into any qualitative analysis. The appendix to this chapter reports mathematical results that we use through the rest of the book.

In this chapter we look at the qualitative analysis of models in which the equilibrium conditions arise from unconstrained and constrained optimization (maximization) processes. We illustrate the application of the results with the traditional models of firm profit maximization and individual utility maximization in the presence of a budget constraint.

In Chapters 2–4 the "signs" of a matrix's entries were often manipulated during the conduct of the analysis. Within this frame of reference, the "qualitative" entries were represented by the mathematical operator "sgn" such that for $a > 0$, sgn $a = 1$, and for $a < 0$, sgn $a = -1$. In this and the remaining chapters, the signs of entries are displayed in the sense of a schematic of the arrays at issue. The analysis itself is based upon other characteristics of the systems being studied. Given this, the signs themselves are displayed in the schematics, i.e., "$+$" to represent $a > 0$ and "$-$" to represent $a < 0$. See note 3 to Chapter 2.

5.2 UNCONSTRAINED MAXIMIZATION

Consider first the case of unconstrained maximization. Let $f(x_1, x_2, \ldots, x_n, \alpha)$ be a real-valued function with continuous second partial derivatives in x and α, where α is a shift parameter. Let

$$f_i = \frac{\partial f}{\partial x_i}, f_{ij} = \frac{\partial^2 f}{\partial x_j \partial x_j}$$

for $i, j = 1, \ldots, n$. To keep things as simple as possible, we follow the lead of Samuelson (1947) in restricting our analysis to the case of "regular" maximum positions, i.e., the case in which second-order conditions hold with strict inequality.

Thus f attains a *regular maximum* at $x^* = (x_1^*, \ldots, x_n^*)$, given that $\alpha = \alpha^*$, if and only if

$$f_i(x^*, \alpha^*) = 0 \; i = 1, \ldots, n \tag{5.1}$$

and

$$\sum_{i=1}^{n} \sum_{j=1}^{n} f_{ij} h_i h_j < 0 \tag{5.2}$$

for any $h = (h_i, \ldots, h_n)$, $h \neq 0$; i.e., the $n \times n$ matrix $[f_{ij}]$ is negative definite, with all derivatives evaluated at (x^*, α^*).

Assume that a change occurs in α. Differentiating (5.1) with respect to α leads to the system

$$\sum_{j=1}^{n} f_{ij}(dx_j^*/d\alpha) = -f_{i\alpha} \; i = 1, \ldots, n, \tag{5.3}$$

where $f_{i\alpha} = \partial^2 f / \partial x_i \partial \alpha$.

In matrix form, (5.3) can be written as

$$Ay = b, \tag{5.4}$$

where $A = [f_{ij}]$, $y = dx^*/d\alpha \equiv [dx_j^*/d\alpha]$, $b = -[f_{i\alpha}]$.

In (5.4), A is a negative definite matrix. Certain questions of interest for comparative statics analysis can now be raised. Given that A is a negative definite matrix and that the sign patterns $(+, -, 0)$ of A and b are known,

For which sign patterns of a negative definite matrix A is A^{-1} of known sign pattern independent of the values of the entries in A?

For which sign patterns of A and b can the sign pattern of $dx^*/d\alpha$ or some of its entries be known independent of the values of the entries in A and b?

The assumption that $f(x, \alpha)$ has continuous second partial derivatives with respect to all variables implies that matrix A is symmetric. It should also be noted that the existence of the inverse matrix A^{-1} is not an issue under the assumptions stated. In particular, A is a symmetric negative definite matrix if and only if A^{-1} is a symmetric negative definite matrix. A useful characterization of negative definite matrices for comparative statics analysis is that for A symmetric, A is negative definite if and only if A is Hicksian, i.e., if and only if every ith-order principal minor of A has sign $(-1)^i$, $i = 1, \dots, n$.

As in previous chapters, Q_A denotes a *qualitative matrix*, i.e., the set of matrices with the same dimensions and sign pattern as A. The subset (perhaps empty) of matrices in Q_A that are symmetric negative definite will be denoted N_A, i.e., $N_A = \{C \mid C \in Q_A$ and C is symmetric negative definite$\}$. Since N_A is nonempty only when A has a negative diagonal and is sign symmetric, i.e., when $\operatorname{sgn} a_{ij} = \operatorname{sgn} a_{ji}$ $i, j = 1, \dots, n$, we will restrict our attention to such matrices.

In terms of this notation, the questions raised are the following:

1. Determine conditions on Q_A such that A has *signed inverse* under the maximization hypotheses, i.e., such that B, $C \in N_A$ implies $B^{-1} \in Q_C^{-1}$.
2. Determine conditions on Q_A, Q_b such that $Ax = b$ is *fully sign solvable* under the maximization hypotheses, i.e., such that $By = c$, $Cz = d$, B, $C \in N_A$, $c, d \in Q_b$ implies $z \in Q_y$.
3. Determine conditions on Q_A, Q_b such that $Ax = b$ is *partially sign solvable* under the maximization hypotheses, i.e., such that some members of x are sign solvable.

Let us turn to the issue of "signing" the inverse matrix in the case of an unconstrained maximum. The assumption that A is symmetric negative definite implies that $\det A$ has sign $(-1)^n$ while principal minors of A of order $n-1$ have sign $(-1)^{n-1}$ so that diagonal entries in A^{-1} are negative. Hence the problem of signing A^{-1} under the unconstrained maximization hypothesis reduces to that of finding conditions on N_A such that off-diagonal entries of the inverses of matrices appearing in N_A are signed. In the arguments that follow, the concept of a dominant diagonal matrix is employed.

Let $A = [a_{ij}]$ be an $n \times n$ real matrix with $a_{ii} < 0$ $i = 1, \dots, n$. Then A is said to be a *dominant diagonal matrix* if

$$|a_{ii}| > \sum_{j \neq i} |a_{ij}| \; i = 1, \dots, n$$

or

$$|a_{ii}| > \sum_{j \neq i} |a_{ji}| \; i = 1, \ldots, n.$$

It is well known (see, e.g., McKenzie 1960), that dominant diagonal matrices with negative diagonal entries are stable matrices. Hence, dominant negative diagonal matrices that are symmetric are negative definite.

As a matter of terminology, the cofactor A_{ij} of the element a_{ij} in A is said to be signed under the maximization hypothesis if $B \in N_A$ implies $\operatorname{sgn} B_{ij} = \operatorname{sgn} A_{ij}$.

Necessary and sufficient conditions for signing off-diagonal entries in A^{-1} under the maximization hypothesis are given in Theorem 5.1.

THEOREM 5.1. *Assume A has negative diagonal and is sign symmetric. The cofactor $A_{ij}(i \neq j)$ of the entry a_{ij} in A is signed under the maximization hypothesis if and only if every path $a(j \to i)$ in A is weakly of the same sign.*

PROOF. For sufficiency, note that under the maximization hypothesis, every principal minor of A of order i has sign $(-1)^i$. By the Maybee determinantal formula (A2.2), every term in the expansion of A_{ij} takes the form of a product of a path $a(j \to i)$ with a principal minor of A. All such terms have weakly the same sign when the paths have weakly the same sign and principal minors of A of order i have sign $(-1)^i$. Hence the condition of the theorem is sufficient for signing B_{ij} under the maximization hypothesis for every $B \in N_A$.

For necessity, assume that there exist two nonzero paths $a^*(j \to i)$ and $a^{**}(j \to i)$ such that $\operatorname{sgn} a^*(j \to i) \neq \operatorname{sgn} a^{**}(j \to i)$. Choose a symmetric $C \in Q_A$ such that all entries in the path $c^*(j \to i)$ are assigned an absolute value of one, and by symmetry, assign the same values to $c^*(i \to j)$. Let $|c_{jj}| = 3$ for $j = 1, \ldots, n$, and assign all other nonzero entries of C an absolute value of ε. Then for ε sufficiently small, C is dominant diagonal, and hence negative definite, so that $C \in Q_A^*$.

Moreover, in C_{ij}, as ε becomes small, the term that dominates the expansion is the product of $c^*(j \to i)$ with diagonal entries, so that the sign of C_{ij} is determined by the sign of $c^*(j \to i)$. In the same way, a matrix $E \in Q_A^*$ can be chosen such that the dominant term in the expansion of E_{ij} is the product of the path in E corresponding to $c^{**}(j \to i)$ with diagonal entries. Since $\operatorname{sgn} c^*(j \to i) \neq \operatorname{sgn} c^{**}(j \to i)$, this implies that $\operatorname{sgn} C_{ij} \neq \operatorname{sgn} E_{ij}$, and hence A_{ij} is not signed under the maximization hypothesis.

An immediate implication of Theorem 5.1 is this:

COROLLARY. *If a_{ij} appears in a negative cycle in A, then A_{ij} is not signed under the maximization hypothesis.*

PROOF. Any negative cycle involving a_{ij} can be written as $a_{ij}a(j \to i) < 0$. Since under the maximization hypothesis, $\operatorname{sgn} a_{ij} = \operatorname{sgn} a_{ji}$, we also have $a_{ji}a(j \to i) < 0$, so that a_{ji} and $a(j \to i)$ are of opposite sign, which violates the condition of Theorem 5.1; hence A_{ij} is not signed under the maximization hypothesis.

In the appendix to this chapter, two classes of matrices of special interest to economists are identified, namely, Metzler matrices and Morishima matrices. A Metzler matrix is a square matrix with all diagonal entries negative and all off-diagonal entries nonnegative. A Morishima matrix is a square matrix with two square diagonal blocks that are Metzler matrices, and with all entries in the off-diagonal blocks non-positive.

The problem of signing A^{-1} is addressed in Theorem 5.2.

THEOREM 5.2. *Assume that A is irreducible, has negative diagonal, and is sign symmetric. Then A has signed inverse under the unconstrained maximization hypothesis if and only if A is either a Metzler matrix or a Morishima matrix.*

PROOF. As discussed in the appendix to this chapter, Metzler and Morishima matrices exhaust the class of irreducible matrices with negative diagonal entries and all cycles nonnegative. The class of Metzler and Morishima matrices that satisfies symmetry and has ith-order principal minors of sign $(-1)^i$, $i = 1, \dots, n$ (i.e., Hicksian matrices) is the class that, in addition, satisfies the maximization hypothesis. As noted in the appendix, Metzler and Morishima matrices that are Hicksian have signed inverses. On the other hand, if A contains a negative cycle, then A^{-1} contains unsigned entries under the maximization hypothesis, from the corollary to Theorem 5.1.

Examples of qualitative matrices and their inverses are the following, where $B \in N_A$. In examples 3 and 4, a question mark indicates an unsigned element in A^{-1} under the maximization hypothesis.

EXAMPLE 1. Metzler matrix.

$$\operatorname{sgn} B = \begin{bmatrix} - & + & + \\ + & - & + \\ + & + & - \end{bmatrix} \quad \operatorname{sgn} B^{-1} = \begin{bmatrix} - & - & - \\ - & - & - \\ - & - & - \end{bmatrix}$$

EXAMPLE 2. Morishima matrix.

$$\operatorname{sgn} B = \begin{bmatrix} - & + & - & - \\ + & - & - & - \\ - & - & - & + \\ - & - & + & - \end{bmatrix} \quad \operatorname{sgn} B^{-1} = \begin{bmatrix} - & - & + & + \\ - & - & + & + \\ + & + & - & - \\ + & + & - & - \end{bmatrix}$$

EXAMPLE 3. Matrix with every index appearing in a negative cycle.

$$\operatorname{sgn} B = \begin{bmatrix} - & - & - \\ - & - & - \\ - & - & - \end{bmatrix} \quad \operatorname{sgn} B^{-1} = \begin{bmatrix} - & ? & ? \\ ? & - & ? \\ ? & ? & - \end{bmatrix}$$

EXAMPLE 4. Matrix with some but not all indices appearing in negative cycles.

$$\operatorname{sgn} B = \begin{bmatrix} - & - & - & 0 & 0 \\ - & - & - & 0 & 0 \\ - & - & - & + & 0 \\ 0 & 0 & + & - & + \\ 0 & 0 & 0 & + & - \end{bmatrix}$$

$$\operatorname{sgn} B^{-1} = \begin{bmatrix} - & ? & ? & ? & ? \\ ? & - & ? & ? & ? \\ ? & ? & - & - & - \\ ? & ? & - & - & - \\ ? & ? & - & - & - \end{bmatrix}$$

Example 1 illustrates the case of a Metzler matrix, which, under the maximization hypothesis, has signed inverse, with all entries in the inverse matrix being negative. In example 2, the case of a Morishima matrix, the signed inverse consists of two diagonal blocks, with all entries negative (the Metzler inverse sign pattern), and off-diagonal blocks with all entries positive. Example 3 is the case in which every off-diagonal element in the matrix appears in a negative cycle so that, by the corollary to Theorem 5.1 all off-diagonal entries in the inverse are unsigned. Finally, example 4 is the case of a matrix in which some but not all indices appear in negative cycles. Off-diagonal entries in rows and columns 1 through 3 are unsigned as in example 3. The remaining unsigned entries also violate the condition that all paths appearing in their cofactors have the same sign.

We next turn to qualitative comparative statics under the (unconstrained) maximization hypothesis. That is, we are interested in determining sign pattern conditions on A and b such that $Ax = b$ is sign solvable under the maximization hypothesis.

The argument to establish necessary and sufficient conditions for full sign solvability is the following. First, we note that if $S = \{1, \ldots, m\}$ denotes the set of indices in A appearing in negative cycles in A, then if $b_i \neq 0$ for some $i \in S$, there cannot be full sign solvability, because the entry of some element in the ith row of the inverse matrix is ambiguously signed under the maximization hypothesis from the corollary to Theorem 5.1. Suppose now that there are both negative and positive cycles in A, with $S' = \{m+1, \ldots, n\}$ denoting the set of indices that appear only in nonnegative cycles in A. If S is nonempty, then Lemma 5.1 states there will *not* be full sign solvability when A is irreducible.

LEMMA 5.1. *Let A be an $n \times n$ irreducible matrix. If A contains both negative and positive cycles, then in A^{-1} there exists an entry in each row and column that is not signed under the maximization hypothesis.*

PROOF. Without loss of generality, assume that the index set of the negative cycles in A is $S = \{1, \ldots, m\}$, with $S' = \{m+1, \ldots, n\}$. By the corollary to Theorem 5.1, for any $i \in S$, there exists an entry in the ith row of A^{-1} that is ambiguously signed under the maximization hypothesis, and similarly for the ith column. By irreducibility of A, there exists $a_{ij} \neq 0$ for some $i \in S$, $j \in S'$. By reindexing, if necessary, take $a_{m,m+1} \neq 0$, and let $a_{rm} a(m \to r)$ denote a negative cycle containing the index m, where $r \in S$. Then, for any such r, consider $A_{r,m+1}$. One nonzero path in $A_{r,m+1}$ is $a_{m+1,m} a_{mr}$, and another is $a_{m+1,m} a(m \to r)$. Since $a_{mr} a(m \to r) < 0$, there exist two nonzero paths in $A_{r,m+1}$ of opposite signs so that $A_{r,m+1} = A_{m+1,r}$ is ambiguously signed under the maximization hypothesis. Consider now any $i \in S'$, $i \neq m+1$. Again by irreducibility, there exists a nonzero path $a(i \to m+1)$. Thus there exist two products, $a(i \to m+1) a_{m+1,m} a(m \to r)$ and $a(i \to m+1) a_{m+1,m} a_{mr}$, of opposite signs. After eliminating cycles, if present, there exist two nonzero paths from i to r with opposite signs so that $A_{ir} = A_{ri}$ is ambiguously signed under the maximization hypothesis. Thus for all $i = 1, \ldots, n$, there exists an entry in the ith row and the ith column of A^{-1} that is ambiguously signed under the maximization hypothesis.

This leads into the following theorem on full sign solvability under the maximization hypothesis.

THEOREM 5.3. *Assume that A is an irreducible matrix that is sign symmetric with negative diagonal entries, and that b is a nonzero vector. Then $Ax = b$ is fully sign solvable under the maximization hypothesis if and only if either*

(i) *all nonzero entries in b are of the same sign and A is a Metzler matrix; or*

(ii) *A is a Morishima matrix with a Metzler submatrix in rows and columns $\{1, \dots, m\}$ and a second Metzler submatrix in rows and columns $\{m+1, \dots, n\}$, and all the nonzero entries in b in rows $\{1, \dots, m\}$ have the same sign, opposite to that of the nonzero entries in rows $\{m+1, \dots, n\}$.*

PROOF. Lemma 5.1 establishes that, for A irreducible, full sign solvability under the maximization hypothesis implies that A has no negative cycles. In the irreducible case, this means that A is either a Metzler matrix or a Morishima matrix.

Under symmetry, stability of A is equivalent to negative definiteness. A necessary and sufficient condition for stability of an irreducible Metzler matrix is that all entries in the inverse matrix be negative. For a Morishima matrix the condition is that diagonal blocks in the inverse corresponding to the Metzler submatrices have all entries negative, with off-diagonal submatrices having all entries positive, this accounting for the sign pattern conditions specified for the b vector.

In general, when both negative and positive cycles appear in A, full sign solvability under the maximization hypothesis occurs if and only if A is reducible, with one diagonal submatrix containing all the negative cycles in A, and with the corresponding entries in the b vector being zero. The second diagonal submatrix and the corresponding b entries obey the conditions of Theorem 5.3.

Examples of fully sign solvable systems under the maximization hypothesis include the following:

(i)

$$\begin{bmatrix} - & + & + \\ + & - & + \\ + & + & - \end{bmatrix} \begin{bmatrix} x_1 \\ x_2 \\ x_3 \end{bmatrix} = \begin{bmatrix} + \\ + \\ + \end{bmatrix} \Rightarrow \begin{bmatrix} x_1 \\ x_2 \\ x_3 \end{bmatrix} = \begin{bmatrix} - & - & - \\ - & - & - \\ - & - & - \end{bmatrix} \begin{bmatrix} + \\ + \\ + \end{bmatrix} = \begin{bmatrix} - \\ - \\ - \end{bmatrix}$$

(ii)

$$\begin{bmatrix} - & + & - & - \\ + & - & - & - \\ - & - & - & + \\ - & - & + & - \end{bmatrix} \begin{bmatrix} x_1 \\ x_2 \\ x_3 \\ x_4 \end{bmatrix} = \begin{bmatrix} + \\ + \\ - \\ - \end{bmatrix}$$

$$\Rightarrow \begin{bmatrix} x_1 \\ x_2 \\ x_3 \\ x_4 \end{bmatrix} = \begin{bmatrix} - & - & + & + \\ - & - & + & + \\ + & + & - & - \\ + & + & - & - \end{bmatrix} \begin{bmatrix} + \\ + \\ - \\ - \end{bmatrix} = \begin{bmatrix} - \\ - \\ + \\ + \end{bmatrix}$$

(iii)

$$\begin{bmatrix} - & - & - & 0 & 0 \\ - & - & - & 0 & 0 \\ - & - & - & 0 & 0 \\ 0 & 0 & 0 & - & + \\ 0 & 0 & 0 & + & - \end{bmatrix} \begin{bmatrix} x_1 \\ x_2 \\ x_3 \\ x_4 \\ x_5 \end{bmatrix} = \begin{bmatrix} 0 \\ 0 \\ 0 \\ + \\ + \end{bmatrix}$$

$$\Rightarrow \begin{bmatrix} x_1 \\ x_2 \\ x_3 \\ x_4 \\ x_5 \end{bmatrix} = \begin{bmatrix} - & ? & ? & 0 & 0 \\ ? & - & ? & 0 & 0 \\ ? & ? & - & 0 & 0 \\ 0 & 0 & 0 & - & - \\ 0 & 0 & 0 & - & - \end{bmatrix} \begin{bmatrix} 0 \\ 0 \\ 0 \\ + \\ + \end{bmatrix} = \begin{bmatrix} 0 \\ 0 \\ 0 \\ - \\ - \end{bmatrix}$$

Example (i) is the Metzler case; (ii) is the Morishima case; and in (iii) the coefficient matrix is reducible, with the Metzler submatrix associated with the nonzero partition of the b vector.

On the other hand, full sign solvability does not hold in this case:

$$\begin{bmatrix} - & - & - & 0 & 0 \\ - & - & - & 0 & 0 \\ - & - & - & + & 0 \\ 0 & 0 & + & - & + \\ 0 & 0 & - & + & - \end{bmatrix} \begin{bmatrix} x_1 \\ x_2 \\ x_3 \\ x_4 \\ x_5 \end{bmatrix} = \begin{bmatrix} 0 \\ 0 \\ 0 \\ + \\ 0 \end{bmatrix}$$

$$\Rightarrow \begin{bmatrix} x_1 \\ x_2 \\ x_3 \\ x_4 \\ x_5 \end{bmatrix} = \begin{bmatrix} - & ? & ? & ? & ? \\ ? & - & ? & ? & ? \\ ? & ? & - & - & ? \\ ? & ? & ? & - & - \\ ? & ? & ? & - & - \end{bmatrix} \begin{bmatrix} 0 \\ 0 \\ 0 \\ + \\ 0 \end{bmatrix} = \begin{bmatrix} ? \\ ? \\ - \\ - \\ - \end{bmatrix}.$$

The matrix A contains negative cycles and is irreducible; hence it violates Theorem 5.3.

Partial sign solvability under the maximization hypothesis occurs for all cases where the b vector has only one nonzero element, since diagonal entries in A^{-1} are signed under the maximization hypothesis. When the b vector contains more than one nonzero entry, there are a number of special cases that can arise. The general rule concerning this is given in the next theorem.

THEOREM 5.4. *Let A be a sign symmetric matrix with negative diagonal entries, with b a nonzero vector. Let $x = (x^1, x^2)$. Then $Ax = b$ is partially* *sign solvable for* $x_i \in x^1$ *if and only if*

(i) $b_i \neq 0$ *is the only nonzero entry in* b; *or*

(ii) *if* $b_j \neq 0$ *for more than one* j, *with* $b_i > 0$, *then every nonzero path* $a(i \to j) > 0$ *for* $b_j > 0$, *with every nonzero path* $a(i \to j) < 0$ *for* $b_j < 0$ (*the signs on the paths are reversed for* $b_i < 0$); *or*

(iii) *if* $b_j \neq 0$ *for one or more entries in* b, *with* $b_i = 0$, *then every nonzero path* $a(i \to j)$ *has the same sign for any such* j, *with all nonzero terms* $A_{ji} b_j$ *of the same sign.*

PROOF. The proof follows immediately from the fact that for $B \in N_A$, diagonal entries in B^{-1} are negative, while off-diagonal entries B_{ji} are signed if and only if every nonzero path $b(i \to j)$ has the same sign, with sgn $B_{ji} = (-)$sgn $b(i \to j)$ for any nonzero path $b(i \to j)$, from the determinantal formula A2.2 of the appendix of Chapter 2.

EXAMPLE 1. $b_1 \neq 0$, $b_i = 0\ i \neq 1$ implies partial sign solvability for x_1.

$$\begin{bmatrix} - & - & - & + \\ - & - & - & + \\ - & - & - & + \\ + & + & + & - \end{bmatrix} \begin{bmatrix} x_1 \\ x_2 \\ x_3 \\ x_4 \end{bmatrix} = \begin{bmatrix} + \\ 0 \\ 0 \\ 0 \end{bmatrix}$$

$$\Rightarrow \begin{bmatrix} x_1 \\ x_2 \\ x_3 \\ x_4 \end{bmatrix} = \begin{bmatrix} - & ? & ? & ? \\ ? & - & ? & ? \\ ? & ? & - & ? \\ ? & ? & ? & - \end{bmatrix} \begin{bmatrix} + \\ 0 \\ 0 \\ 0 \end{bmatrix} = \begin{bmatrix} - \\ ? \\ ? \\ ? \end{bmatrix}$$

EXAMPLE 2. $b_1 \neq 0$, $b_2 \neq 0$, sgn $b_1 =$ sgn b_2. Full sign solvability (and hence partial sign solvability) in the Metzler case.

$$\begin{bmatrix} - & + & + & + \\ + & - & + & + \\ + & + & - & + \\ + & + & + & - \end{bmatrix} \begin{bmatrix} x_1 \\ x_2 \\ x_3 \\ x_4 \end{bmatrix} = \begin{bmatrix} + \\ + \\ 0 \\ 0 \end{bmatrix}$$

$$\Rightarrow \begin{bmatrix} x_1 \\ x_2 \\ x_3 \\ x_4 \end{bmatrix} = \begin{bmatrix} - & - & - & - \\ - & - & - & - \\ - & - & - & - \\ - & - & - & - \end{bmatrix} \begin{bmatrix} + \\ + \\ 0 \\ 0 \end{bmatrix} = \begin{bmatrix} - \\ - \\ - \\ - \end{bmatrix}$$

EXAMPLE 3. $b_i \neq 0\ i = 1, \ldots, 4$, sgn $b_1 =$ sgn b_2, sgn $b_3 =$ sgn b_4, sgn $b_1 \neq$ sgn b_3. Full and partial sign solvability in the Morishima case.

$$\begin{bmatrix} - & + & - & - \\ + & - & - & - \\ - & - & - & + \\ - & - & + & - \end{bmatrix} \begin{bmatrix} x_1 \\ x_2 \\ x_3 \\ x_4 \end{bmatrix} = \begin{bmatrix} + \\ + \\ - \\ - \end{bmatrix}$$

$$\Rightarrow \begin{bmatrix} x_1 \\ x_2 \\ x_3 \\ x_4 \end{bmatrix} = \begin{bmatrix} - & - & + & + \\ - & - & + & + \\ + & + & - & - \\ + & + & - & - \end{bmatrix} \begin{bmatrix} + \\ + \\ - \\ - \end{bmatrix} = \begin{bmatrix} - \\ - \\ + \\ + \end{bmatrix}$$

EXAMPLE 4. $b_i \neq 0$ for all i. Partial (but not full) sign solvability in a matrix with negative and positive cycles.

$$\begin{bmatrix} - & + & 0 & 0 & 0 \\ + & - & 0 & 0 & 0 \\ 0 & 0 & - & - & - \\ 0 & 0 & - & - & - \\ 0 & 0 & - & - & - \end{bmatrix} \begin{bmatrix} x_1 \\ x_2 \\ x_3 \\ x_4 \\ x_5 \end{bmatrix} = \begin{bmatrix} + \\ + \\ + \\ + \\ + \end{bmatrix}$$

$$\Rightarrow \begin{bmatrix} x_1 \\ x_2 \\ x_3 \\ x_4 \\ x_5 \end{bmatrix} = \begin{bmatrix} - & - & 0 & 0 & 0 \\ - & - & 0 & 0 & 0 \\ 0 & 0 & - & ? & ? \\ 0 & 0 & ? & - & ? \\ 0 & 0 & ? & ? & - \end{bmatrix} \begin{bmatrix} + \\ + \\ + \\ + \\ + \end{bmatrix} = \begin{bmatrix} - \\ - \\ ? \\ ? \\ ? \end{bmatrix}$$

EXAMPLE 5. $b_1 \neq 0$, $b_2 \neq 0$, $b_3 \neq 0$. Partial (but not full) sign solvability in an irreducible matrix with negative and positive cycles.

$$\begin{bmatrix} - & + & 0 & 0 & 0 \\ + & - & + & 0 & 0 \\ 0 & + & - & - & - \\ 0 & 0 & - & - & - \\ 0 & 0 & - & - & - \end{bmatrix} \begin{bmatrix} x_1 \\ x_2 \\ x_3 \\ x_4 \\ x_5 \end{bmatrix} = \begin{bmatrix} + \\ + \\ + \\ 0 \\ 0 \end{bmatrix}$$

$$\Rightarrow \begin{bmatrix} x_1 \\ x_2 \\ x_3 \\ x_4 \\ x_5 \end{bmatrix} = \begin{bmatrix} - & - & - & ? & ? \\ - & - & - & ? & ? \\ - & - & - & ? & ? \\ ? & ? & ? & - & ? \\ ? & ? & ? & ? & - \end{bmatrix} \begin{bmatrix} + \\ + \\ + \\ 0 \\ 0 \end{bmatrix} = \begin{bmatrix} - \\ - \\ - \\ ? \\ ? \end{bmatrix}$$

It is easy to verify that there is partial sign solvability in the Metzler and Morishima cases (Examples 2 and 3) if and only if there is full sign solvability.

Milgrom and Shannon (1994) have derived a necessary and sufficient condition for the solution set of an optimization problem to be monotonic in the parameters of the problem. Their main theorem (theorem 4) states that the solution set $x(t)$ of an optimization problem with objective function $f(x, t)$ is monotone nondecreasing with respect to a change in the parameters t of the problem if and only if f is quasi-supermodular in x and satisfies the single crossing property in $(x; t)$. Theorems 5 and 6 of Milgrom and Shannon then specialize this result to the case where f is assumed to be twice continuously differentiable. In this case, and with x an n-vector and t an m-vector, the Milgrom-Shannon necessary and sufficient conditions for $x(t)$ to be monotone nondecreasing in t are that the Jacobian matrix be Metzler and the b vector be nonnegative and nonzero.

Theorem 5.3 can be applied to obtain a similar result for the class of qualitatively specified maximization models, while at the same time extending that result to identify another basic comparative statics result. In the first place, Theorem 5.3 identifies the Metzler and Morishima cases as the only qualitatively specified cases in which full sign solvability (complete comparative statics results) can be obtained concerning the direction of change in the equilibrium values of all endogenous variables. Clearly, in the Metzler case, with all elements in the b vector weakly of the same sign, every element in the solution vector is positive (under irreducibility), so that the equilibrium values are monotone increasing in the parameter values. In the Morishima case (see example (ii) in the enumeration of fully sign solvable systems above), every element in the solution vector is signed, so that there are definite qualitative comparative statics results in this case. However, the signs of the terms with indices in the index set of the first Metzler block of the Morishima matrix differ from those of the terms with indices in the index set of the second Metzler block, so that only a subset of the endogenous variables are monotone increasing in the parameter while the remainder are monotone decreasing.

The use of a specific qualitative framework has permitted the derivation of results beyond those contained in the Milgrom-Shannon paper. No doubt, this simply reflects the fact that the emphasis in the approach taken here is precisely on identifying conditions for full and partial sign solvability, while Milgrom and Shannon are instead interested in showing that their approach, while centered on problems relating to the existence of comparative statics results, also can be utilized in applications.

5.3 APPLICATION: PROFIT-MAXIMIZING CHOICES OF INPUTS BY A COMPETITIVE FIRM

Consider the traditional comparative statics problem involving the analysis of the profit-maximizing input and output choices of a competitive firm. Let y denote the number of units of output, p the price per unit of output, x_i the number of units of the ith input hired, and w_i the price per unit of the ith input. We assume the existence of a strictly concave production function, $y = f(x_1, \ldots, x_n)$, where output is a strictly increasing function of each input. The firm chooses amounts of inputs to maximize profit, π, where

$$\pi = py - \sum_{i=1}^{n} w_i x_i.$$

The equilibrium conditions characterizing a regular maximum of profits are

(1) $\partial\pi/\partial x_i = pf_i - w_i = 0 \quad i = 1, \ldots, n$; and
(2) $[pf_{ij}]$ is a symmetric negative definite matrix,

where $f_i = \partial f/\partial x_i$, $f_{ij} = \partial^2 f/\partial x_i x_j$ $i, j = 1, \ldots, m$.

Differentiating condition (1) totally with respect to $p, w_1, \ldots, w_n$, we obtain

$$[pf_{ij}] \begin{bmatrix} dx_1 \\ \vdots \\ dx_n \end{bmatrix} = \begin{bmatrix} dw_1 - f_1 dp \\ \vdots \\ dw_n - f_n dp \end{bmatrix}.$$

Assume now that the signs $(+, -, 0)$ of the entries in $[pf_{ij}]$ are specified; i.e., assume that the complementarity or substitutability relations among inputs are specified, but not the quantitative magnitudes of such relations. For which such sign patterns is it possible to solve for the signs of all $\partial x_i/\partial p$ ($dw_i = 0$ for $i = 1, \ldots, n$), and for the signs of all $\partial x_i/\partial w_j$ ($dp = 0$ and $dw_k = 0$ for $k \neq j$)? We consider the case in which $[pf_{ij}]$ is irreducible. From Lemma 5.1 and Theorem 5.3 of the previous section, we have the following results:

1. Only in the case in which $f_{ij} \geq 0$ for all $i \neq j$ (all inputs are complements) is it possible to solve for the signs of all $\partial x_i/\partial p$, since it is only in this case that the entire inverse matrix is signed, with all entries in the inverse of the same sign. Because $f_i > 0$ for $i = 1, \ldots, n$, every term in the inverse must be weakly of the same

sign if complete comparative statics information is to be present. When all inputs are complements, an increase in the price of the product will lead to an increase in the quantity demanded of *every* input; i.e., $\partial x_i/\partial p > 0$ for $i = 1, \ldots, n$.

2. When $dp = 0$ and $dw_k = 0$ for $k \neq j$, complete comparative statics information is available concerning the signs of $\partial x_i/\partial w_j$ ($i = 1, \ldots, n$) from the maximization hypothesis and qualitative information in only two cases: the Metzler case and the Morishima case. In the Metzler case, in which all inputs are complements, the inverse matrix consists solely of negative entries, so that an increase in w_j leads to a decrease in the quantity demanded of each input (and hence a decrease in the quantity of output supplied as well).

In the Morishima case, the matrix $[pf_{ij}]$ can be written as

$$[pf_{ij}] = \left[\begin{array}{c|c} M_{11} & M_{12} \\ \hline M_{21} & M_{22} \end{array}\right],$$

with M_{11} and M_{22} Metzler, and M_{12} and M_{21} having all nonpositive entries. The matrix's corresponding inverse sign pattern is

$$\left[\begin{array}{c|c} - & + \\ \hline + & - \end{array}\right].$$

Then an increase in w_j leads to a decrease in the quantity demanded of input i if row i appears in M_{11} or M_{22} ($f_{ij} \geq 0$ $i \neq j$), i.e., if i and j are complements, and to an increase in the quantity demanded of input i if row i appears in M_{12} or M_{21} ($f_{ij} \leq 0$ $i \neq j$), i.e., if i and j are substitutes in production. An increase in w_j always leads to a decrease in the quantity demanded of input j, of course, since diagonal entries in the inverse matrix are negative.

Partial sign solvability occurs in this model in connection with changes in w_j, with other input prices, and the price of output being held constant. When only w_j changes, then under the maximization hypothesis, the quantity demanded of input j decreases, whatever the sign pattern of $[pf_{ij}]$ (this is the case of a single nonzero entry in the b vector).

Further, if there is a Metzlerian block in the coefficient matrix (as in examples 4 and 5 in the discussion accompanying Theorem 5.4 above), predictable changes in the quantities demanded of inputs other than j can be derived as well, even for irreducible systems.

Finally, we might note one well-known important comparative statics result that holds independently of the sign pattern of $[pf_{ij}]$, namely, homogeneity of degree zero in $(p, w_1, \ldots, w_n)$ of input demands and the supply of output; i.e., an equal proportionate increase in input and output prices leads to no change in input hirings or the quantity of output supplied. This is not a qualitative result, since the "equal proportionate increase" assumption involves quantitative information.

5.4 CONSTRAINED MAXIMIZATION

Consider the following problem: Choose $y = (y_1, \ldots, y_n)$ to maximize $f(y, \alpha)$ subject to $g(y, \alpha) = 0$, where α is a shift parameter.

Let $L = f + \lambda g$ denote the Lagrangian expression, where λ is the Lagrange multiplier. Suppose that a constrained maximum occurs at y^*. First-order conditions are given by

$$L_i(y^*, \alpha) = f_i(y^*, \alpha) + \lambda g_i(y^*, \alpha) = 0 \; i = 1, \ldots, n; \qquad (5.5)$$
$$L_\lambda(y^*, \alpha) = g(y^*, \alpha) = 0.$$

A regular constrained maximum occurs at y^* if the first-order conditions are satisfied and the following second-order condition holds:

$$\sum_{i=1}^{n} \sum_{j=1}^{n} L_{ij}(y^*, \alpha) h_i h_j < 0, \text{ for } h = (h_1, \ldots, h_n) \neq 0,$$
$$\text{subject to } \sum_{i=1}^{n} g_i(y^*, \alpha) h_i = 0.$$

Let

$$\hat{A} = \begin{bmatrix} A & a \\ a' & 0 \end{bmatrix},$$

where $A = [L_{ij}]$ is an $n \times n$ matrix, $a = [g_i]$ is an $n \times 1$ vector, and a' is the transpose of a. Then the second-order conditions above are satisfied if the bordered principal minors of A alternate in sign, with any $r \times r$ bordered principal minor having sign $(-1)^{r+1}$.

Let

$$x = \begin{bmatrix} y^* \\ \lambda \end{bmatrix}$$

denote an $(n+1)\times 1$ vector. The comparative statics problem we are interested in involves determining the effect on the equilibrium vector x of changes in the shift parameter α. Differentiating the first-order conditions with respect to α, we obtain a system of $n+1$ equations in $n+1$ variables (the n values of the ys plus the Lagrange multiplier, λ), which appear as follows:

$$\sum_{j=1}^{n} (f_{ij} + \lambda g_{ij}) dy_j + g_i d\lambda = -[f_{i\alpha} + \lambda g_{i\alpha}] d\alpha; \tag{5.6}$$

$$\sum_{j=1}^{n} g_j dy_j = -g_i d\alpha,$$

where $f_{i\alpha} = \partial^2 f/\partial y_i \partial\alpha$, $g_{i\alpha} = \partial^2 g/\partial y_1 \partial\alpha$, and $g_\alpha = \partial g/\partial\alpha$.

In matrix form, (5.6) can be written as

$$\hat{A}x = b, \tag{5.7}$$

where $\hat{A}$ is the bordered matrix above, x is the $(n+1)\times 1$ vector of derivatives of y and λ with respect to α, and $(-)b$ is the $n+1\times 1$ vector of partial derivatives of the equilibrium conditions with respect to α.

At a regular constrained maximum, $\hat{A}^{-1}$ exists, with sgn det $\hat{A} = (-1)^n$ by the second-order conditions. When $\hat{A}^{-1}$ is written in partitioned form, as

$$\hat{A}^{-1} = \left[\begin{array}{c|c} C & c \\ \hline c' & d \end{array}\right],$$

it is well known that C is a matrix negative semidefinite of rank $n-1$ (see, e.g., the Mathematical Appendix of Samuelson [1947]).

Let $M_{\hat{A}} = \{\{\hat{B} \mid \hat{B} \in Q_{\hat{A}}$ and $\hat{B}$ satisfies the bordered principal minor conditions$\}$. Formally, we say that $\hat{A}x = b$ is sign solvable under the constrained maximization hypothesis if $\hat{B} \in M_{\hat{A}}$, $d \in Q_b$, and $\hat{B}z = d$ implies that $z \in Q_x$; and there is partial sign solvability under the constrained maximization hypothesis if there is sign solvability under that hypothesis for some members of x. As in the case of an unconstrained maximum, M_A is empty except when $\hat{A}$ is sign symmetric, which is the only case we consider.

An example of a system sign solvable under constrained maximization is the following:

$$\begin{bmatrix} - & + & + \\ + & - & + \\ + & + & 0 \end{bmatrix}\begin{bmatrix} x_1 \\ x_2 \\ x_3 \end{bmatrix} = \begin{bmatrix} + \\ 0 \\ 0 \end{bmatrix} \Rightarrow \begin{bmatrix} x_1 \\ x_2 \\ x_3 \end{bmatrix} = \begin{bmatrix} - & + & + \\ + & - & + \\ + & + & ? \end{bmatrix}\begin{bmatrix} + \\ 0 \\ 0 \end{bmatrix} = \begin{bmatrix} - \\ + \\ + \end{bmatrix}.$$

In what follows, we will take as our "standard case" that in which all entries in the $a(a')$ vector are positive and diagonal entries in A are negative. While this is not the most general case, it simplifies the notation considerably, and covers most cases of interest in economic theory. The leading instance of this case for sign solvability purposes is that in which all off-diagonal entries in A are zero. It is easy to verify that in this case, all entries in the inverse matrix are positive, except for the first n diagonal entries, which are negative. Note that even in this special case, full sign solvability occurs only if the b vector contains only one nonzero entry. Since adding nonzero off-diagonal entries can only add ambiguous entries in the inverse, this means that a necessary condition for full sign solvability under the constrained maximization hypothesis is that $b_k \neq 0$ for at most one index k.

Except for the case $n = 2$, depicted in the example above (recall that the coefficient matrix is $(n+1) \times (n+1)$), a nonzero off-diagonal element a_{ij} implies that all cofactors of $\hat{A}$ that include rows and columns i and j are unsigned under the unconstrained maximization hypothesis.

LEMMA 5.2. *Let $\hat{A}$ be sign symmetric and in standard form; i.e., let diagonal entries in A be negative, and entries in the bordering row and column of $\hat{A}$ be positive, with $n > 2$. Then if there exists a nonzero entry $a_{ij}(= a_{ji})$, $i \neq j$, $i, j = 1, \dots, n$, this implies that $\hat{A}_{rs}$, $r \neq s$ is ambiguously signed under the unconstrained maximization hypothesis, for $r, s \neq i$ and $r, s \neq j$, $r, s = 1, \dots, n+1$; and $\hat{A}_{n+1,n+1}$ is ambiguously signed as well.*

PROOF. We sketch the proof for $\hat{A}_{n+1,n+1}$; the same kind of argument applies for the other cofactors. Without loss of generality, assume that $a_{12} \neq 0$ (which implies $a_{21} \neq 0$). Let $D \in Q_A$ be chosen so that $d_{11} = -3$ $\cdot d_{ii} = -1$ $i = 2, \dots, n$, $|d_{12}| = |d_{21}| = 2$. In $\hat{D}$, let d denote the vector bordering D, and choose $d_1 = 1$, $d_2 = \varepsilon$. Let all other nonzero entries in $\hat{D}$ be chosen with absolute value ε. Then it is easy to verify that bordered principal minors of $\hat{D}$ of dimension $r \times r$ have sign $(-1)^{r-1}$, so that $\hat{D} \in M_{\hat{A}}$, while sgn $\hat{D}_{n+1,n+1} = (-1)^{n+1}$. On the other hand, let $\hat{C} \in Q_{\hat{M}}$ be chosen with all diagonal entries in C having value (-1), with all other nonzero entries in C with absolute value ε, and with entries in the bordering c and c' vectors equal to 1. Then $\hat{C} \in M_{\hat{A}}$ and $\operatorname{sgn} \hat{C}_{n+1,n+1} = (-1)^n$ so that $\hat{A}_{n+1,n+1}$ is not signed.

LEMMA 5.3. *Let $\hat{A}$ be sign symmetric in standard form, with $n > 2$. Then if there exists a nonzero entry a_{ij} $i \neq j$, $i, j = 1, \ldots, n$, $a_{ij} > 0$ implies $\hat{A}_{ij}$ is not signed under constrained maximization, while $a_{ij} < 0$ implies that $\hat{A}_{ik}$ and $\hat{A}_{jk}$ are not signed, for $k \pm i$, $j, k = 1, \ldots, n+1$.*

PROOF. If $a_{ij} > 0$, then even in the case where all other off-diagonal entries in A are zero, there are two terms in the expansion of $\hat{A}_{ij}$ of opposite sign, namely, a_{ji} multiplied by a bordered principal minor of dimensions $(n-1) \times (n-1)$, and a second term, $(-)a_j^2$, multiplied by an $(n-2)$–order principal minor of A. The bordered principal minor and the $(n-2)$–order principal minor both have sign $(-1)^{n-2}$, and either can be chosen arbitrarily large relative to the other without violating the constrained maximization hypothesis. Thus there exist $\hat{B}$, $\hat{D} \in M_A$ with $\operatorname{sgn} \hat{B}_{ij} \neq \operatorname{sgn} \hat{D}_{ij}$. Hence $a_{ij} > 0$ implies $\hat{A}_{ij}$ is not signed.

If $a_{ij} < 0$, then, similarly, there will be two terms of opposite sign in each off-diagonal cofactor $\hat{A}_{ik}$, $\hat{A}_{jk}$, $k \neq i$, $j, k = 1, \ldots, m+1$, even in the case in which all other off-diagonal terms in A are zero. It is easy to verify that each can be chosen arbitrarily large relative to the other without violating the constrained maximization hypothesis, so that $a_{ij} < 0$ implies that $\hat{A}_{ik}$, $\hat{A}_{jk}$ is not signed for $k \neq i$, $j, k = 1, \ldots, n+1$. This leads into the following theorems on sign solvability under the constrained maximization hypothesis.

THEOREM 5.5. *Let $\hat{A}$ be an $(n+1) \times (n+1)$ sign symmetric matrix in standard form, $n > 2$. Then $\hat{A}x = b$ is fully sign solvable under the constrained maximization hypothesis if and only if $b_k \neq 0$ for at most one index $k \in \{1, \ldots, n+1\}$ and $a_{ij} = 0$ for every $i \neq j$.*

PROOF. As noted earlier, full sign solvability implies that the b vector contains at most one nonzero entry. Lemmas 5.2 and 5.3 imply that full sign solvability requires that all off-diagonal entries in A be zero, and when A is a diagonal matrix, all of the entries in $\hat{A}^{-1}$ are signed under the constrained maximization hypothesis, so there is full sign solvability.

THEOREM 5.6. *Let $\hat{A}$ be an $(n+1) \times (n+1)$ sign symmetric matrix in standard form, $n > 2$. Let $b = \{b^1, b^2, b^3\}$, where b^1 contains all positive entries in b, b^2 contains all negative entries in b, and b^3 contains all zero entries in b. Then there is partial sign solvability under the constrained maximization hypothesis for x_k, $k = 1, \ldots, n$, if and only if one of the following conditions holds:*

(i) $b_k \neq 0$ *is the unique non-zero entry in* b; *and/or*

(ii) $a_{ij} = 0$ *for* $i \neq j, i, j = 1, \ldots, n$, *and* $b_k \neq 0$ *is the unique non-zero entry in either* b^1 *or* b^2; *and/or*

(iii) $a_{ik} \neq 0 a_{ki} \neq 0$ *for some* $i \neq k$ *are the only non-zero off diagonal entries in* A, *where* $a_{ik} > 0$ *implies* $b_i = b_k = 0$, *and either* $b^1 = \varnothing$ *or* $b^2 = \varnothing$, *while* $a_{ik} < 0$ *implies* $\operatorname{sgn} b_i \neq \operatorname{sgn} b_k$, *and* $b_j \in b^3$ *for all* $j \neq k, i$.

Finally, under the conditions of the theorem, there is sign solvability for x_{n+1} *if and only if* $a_{ij} = 0$ $i \neq j, i, j = 1, \ldots, n$, *and either* $b^1 = \varnothing$ *or* $b^2 = \varnothing$.

PROOF. The necessity of the above conditions follows directly from Lemmas 5.2 and 5.3. Sufficiency of (i) follows because under the constrained maximization hypothesis, the first n diagonal entries in $\hat{A}^{-1}$ are negative, so that x_k is signed. Under (ii), all off-diagonal entries in $\hat{A}^{-1}$ are positive, while the first n diagonal entries are negative. If b_k is the only positive entry in b, then $(\hat{A}_{jk}/\det \hat{A})b_j \leq 0$ $j = 1, \ldots, n+1$; similarly, if b_k is the only negative entry in b, then $(\hat{A}_{jk}/\det \hat{A})b_j \geq 0$ $j = 1, \ldots, n+1$. In either case x_k is signed.

Under (iii) if $a_{ik} = a_{ki} > 0$, then $\hat{A}_{ik} = \hat{A}_{ki}$ are not signed, while all other off-diagonal entries $(\hat{A}_{rs}/\det \hat{A})$ of $\hat{A}^{-1}$ are positive. Hence $b_i = b_k = 0$, with $b_j \geq 0$ $j \neq i, j \neq k$ implies x_k is signed, and similarly if $b_j \leq 0$ $j \neq i$, $j \neq k$. If $a_{ik} = a_{ki} < 0$, then $(\hat{A}_{ik}/\det \hat{A}) = (\hat{A}_{ki}/\det \hat{A}) > 0$, while all other off-diagonal entries in $\hat{A}^{-1}$ are not signed. Hence $\operatorname{sgn} b_i \neq \operatorname{sgn} b_k$, with $b_j = 0$ $j \neq i$, $j \neq k$ implies x_k is signed. (Note that partial sign solvability for several entries of the x vector requires that the above conditions be satisfied for each of the sign solvable entries of the x vector.)

Examples of the sign solvability theorems are the following:

EXAMPLE 1. Full sign solvability.

$$\begin{bmatrix} - & 0 & 0 & 0 & + \\ 0 & - & 0 & 0 & + \\ 0 & 0 & - & 0 & + \\ 0 & 0 & 0 & - & + \\ + & + & + & + & 0 \end{bmatrix} \begin{bmatrix} x_1 \\ x_2 \\ x_3 \\ x_4 \\ x_5 \end{bmatrix} = \begin{bmatrix} + \\ 0 \\ 0 \\ 0 \\ 0 \end{bmatrix}$$

$$\Rightarrow \begin{bmatrix} x_1 \\ x_2 \\ x_3 \\ x_4 \\ x_5 \end{bmatrix} = \begin{bmatrix} - & + & + & + & + \\ + & - & + & + & + \\ + & + & - & + & + \\ + & + & + & - & + \\ + & + & + & + & + \end{bmatrix} \begin{bmatrix} + \\ 0 \\ 0 \\ 0 \\ 0 \end{bmatrix} = \begin{bmatrix} - \\ + \\ + \\ + \\ + \end{bmatrix}$$

EXAMPLE 2. Partial sign solvability for x_1 and x_2.

$$\begin{bmatrix} - & 0 & 0 & 0 & + \\ 0 & - & 0 & 0 & + \\ 0 & 0 & - & 0 & + \\ 0 & 0 & 0 & - & + \\ + & + & + & + & 0 \end{bmatrix} \begin{bmatrix} x_1 \\ x_2 \\ x_3 \\ x_4 \\ x_5 \end{bmatrix} = \begin{bmatrix} + \\ - \\ 0 \\ 0 \\ 0 \end{bmatrix}$$

$$\Rightarrow \begin{bmatrix} x_1 \\ x_2 \\ x_3 \\ x_4 \\ x_5 \end{bmatrix} = \begin{bmatrix} - & + & + & + & + \\ + & - & + & + & + \\ + & + & - & + & + \\ + & + & + & - & + \\ + & + & + & + & + \end{bmatrix} \begin{bmatrix} + \\ - \\ 0 \\ 0 \\ 0 \end{bmatrix} = \begin{bmatrix} - \\ + \\ ? \\ ? \\ ? \end{bmatrix}$$

EXAMPLE 3. Partial sign solvability.

$$\begin{bmatrix} - & - & 0 & 0 & + \\ - & - & 0 & 0 & + \\ 0 & 0 & - & 0 & + \\ 0 & 0 & 0 & - & + \\ + & + & + & + & 0 \end{bmatrix} \begin{bmatrix} x_1 \\ x_2 \\ x_3 \\ x_4 \\ x_5 \end{bmatrix} = \begin{bmatrix} + \\ - \\ 0 \\ 0 \\ 0 \end{bmatrix}$$

$$\Rightarrow \begin{bmatrix} x_1 \\ x_2 \\ x_3 \\ x_4 \\ x_5 \end{bmatrix} = \begin{bmatrix} - & + & ? & ? & ? \\ + & - & ? & ? & ? \\ ? & ? & - & ? & ? \\ ? & ? & ? & - & ? \\ ? & ? & ? & ? & ? \end{bmatrix} \begin{bmatrix} + \\ - \\ 0 \\ 0 \\ 0 \end{bmatrix} = \begin{bmatrix} - \\ + \\ ? \\ ? \\ ? \end{bmatrix}$$

5.5 APPLICATION: MINIMIZING THE COST OF PRODUCTION

In production theory, the problem of maximizing output for a given level of cost (or minimizing cost for a given level of output) illustrates Theorems 5.5 and 5.6 of the preceding section. Let z_i denote the number of units of the ith input, let w_i denote the per-unit price of the ith input, and let $y = f(z_1, \ldots, z_n)$ denote output, with $f_i > 0$ for every i.

The conditions for a regular constrained maximum of output are

(1) $$f_i + \lambda w_i = 0 \qquad i = 1, \dots, n;$$

(2) $$\sum_{i=1}^{n} w_i z_i = c; \text{ and}$$

(3) $$\hat{A} = \left[\begin{array}{c|c} f_{ij} & w_i \\ \hline w_j & 0 \end{array}\right] \text{ is negative definite under constraint,}$$

where c is the given level of cost and $-1/\lambda$ is marginal cost.

From the total differentials of the first-order conditions,

$$[\hat{A}]\begin{bmatrix} dz_1 \\ \vdots \\ dz_n \\ \hline d\lambda \end{bmatrix} = -\begin{bmatrix} \lambda dw_1 \\ \vdots \\ \lambda dw_n \\ \sum z_i \, dw_i - dc \end{bmatrix}.$$

If $n > 2$, and assuming $dw_i = 0$, $i = 1, \dots, m$, the signs of all $\frac{\partial z_i}{\partial c}$ can be obtained only in the case of "complete independence," where $f_{ij} = 0$ for all $i \neq j$. As indicated by example 1 above, the inverse matrix in this case has the sign pattern

$$\begin{bmatrix} - & + & + & \cdots & + & + \\ + & - & + & \cdots & + & + \\ + & + & - & \cdots & + & + \\ \cdots & \cdots & \cdots & \cdots & \cdots & \cdots \\ + & + & + & \cdots & - & + \\ + & + & + & \cdots & + & + \end{bmatrix},$$

while with $dw_i = 0$, $i = 1, \dots, n$, the only nonzero entry in the right-hand vector is $+1$ in the $n+1$st position.

Hence, under "complete independence," $\frac{\partial z_i}{\partial c} > 0$ for every input i, and $\frac{\partial \lambda}{\partial c} > 0$ as well (an increase in allowable cost increases the use of all inputs, and results in an increase in marginal cost as well). For a change in the jth wage rate, w_j, with other wage rates and allowable cost fixed, the right-hand vector has $-\lambda$ in the jth position, $-z_j$ in the $n+1$st position, and zeros elsewhere. Since $\lambda < 0$, $z_j > 0$, this means that there are positive entries in the jth and $n+1$st positions in the right-hand vector. Thus the only signed entry in the solution vector, even under the "complete independence" assumption, is $\frac{\partial z_j}{\partial w_j} < 0$; the signs of all other entries $\frac{\partial z_j}{\partial w_j}$ $i \neq j$, and $\frac{\partial \lambda}{\partial w_j}$, are ambiguous, given only qualitative information coupled with the constrained maximization hypothesis.

APPENDIX: ADVANCED TOPICS IN MATRIX ANALYSIS

A5.1 The Perron-Frobenius Theorem

We first consider square matrices all of whose entries are nonnegative. We use the following notations:

$$A \geq B \text{ if } a_{ij} \geq b_{ij} \text{ for all } i \text{ and } j;$$

$$A > B \text{ if } A \geq B \text{ and } A \neq B;$$

and

$$A \gg B \text{ if } a_{ij} > b_{ij} \text{ for all } i \text{ and } j.$$

The matrices A satisfying $A \gg 0$ are positive matrices.

Our first result is the Perron-Frobenius theorem, proved for positive matrices by Perron in 1907 and for nonnegative irreducible matrices by Frobenius in 1912.

THEOREM A5.1 (PERRON-FROBENIUS). *Let A be an irreducible nonnegative matrix. Then A has an eigenvalue λ^* that is simple (not repeated), positive, with $\lambda^* \geq |\lambda|$ for any other eigenvalue; and if $|\lambda| = \lambda^*$, then λ is also a simple eigenvalue. Moreover, A has a strictly positive eigenvector x^* associated with λ^*.*

The most thorough account of the Perron-Frobenius theorem and its close relatives appears in Berman and Plemmons (1979). We refer the reader to that book for proofs of Theorem A5.1 and subsequent related results.

COROLLARY. (a) *If $0 \leq A \leq B$, then $\lambda^*(A) \leq \lambda^*(B)$.*

(b) *If $0 \leq A < B$ and $A + B$ is irreducible, then $\lambda^*(A) < \lambda^*(B)$.*

(c) *If B is a principal submatrix of $A \geq 0$, then $\lambda^*(B) \leq \lambda^*(A)$.*

(d) *$\lambda^*(A)$ is an eigenvalue of some proper principal submatrix of a nonnegative matrix A if and only if A is reducible.*

We now define the nonnegative matrix A to be *primitive* if there exists a positive integer m such that A^m is positive.

THEOREM A5.2. *If $A \geq 0$ and primitive, then $\lambda^*(A) > |\lambda|$ for any other eigenvalue λ of A.*

If A is primitive and s is a positive integer, then A^T and A^s are irreducible and primitive. We have the following max/min and min/max characterization for the maximal eigenvalue:

$$\lambda^*(A) = \max_{x>0} \left\{ \min_{x_i>0} \frac{(Ax)_i}{x_i} \right\},$$

and

$$\lambda^*(A) = \min_{x>0} \left\{ \max_{x_i>0} \frac{(Ax)_i}{x_i} \right\}.$$

The result presented in Theorem A5.1 is only the first part of the Perron-Frobenius theorem. We now state the second part of the Theorem.

THEOREM A.5.3. (a) *If $A \geq 0$, irreducible, and has eigenvalues $\lambda_0 = \lambda^*(A)$, $\lambda_1 = \lambda^*(A)e^{i\theta_1}, \ldots, \lambda_{h1} = \lambda^*(A)e^{i\theta_{h-1}}$ of modulus $\lambda^*(A)$, $0 < \theta_1 < \theta_2, < \cdots < \theta_{n-1} < 2\pi$, then these numbers are distinct roots of $\lambda^h - (\lambda^*(A))^h = 0$.*

(b) *The entire spectrum $\sigma(A)$ of A goes over into itself under a rotation of the complex plane by π/h.*

(c) *If $h > 1$, then there exists a permutation matrix P such that*

$$PAP^T = \begin{bmatrix} 0 & A_{12} & 0 & \cdots & 0 \\ 0 & 0 & A_{23} & \cdots & 0 \\ \vdots & \vdots & \vdots & \cdots & \vdots \\ 0 & 0 & 0 & \cdots & A_{h-1,h} \\ A_{h1} & 0 & 0 & \cdots & 0 \end{bmatrix},$$

where the zero submatrices along the diagonal are square.

There are many generalizations of the Perron-Frobenius theorem both in the finite-dimensional case and in the infinite-dimensional case. For the finite-dimensional linear case, the reader should consult Berman and Plemmons (1979).

A5.2 Matrices Sign Similar to Nonnegative Matrices

We remind the reader that a signature matrix A is a diagonal matrix $S = \text{diag}[d_1, d_2, \ldots, d_n]$, where each d_i is either 1 or -1. Of course, $\det S = \pm 1$ for all such matrices; hence every such matrix is nonsingular. If $\det S = 1$ and $d_i = 1$, then the cofactors $\prod_{j \neq i} d_j = 1$, and hence the ith diagonal entry of D^{-1} equals 1. On the other hand, if $d_i = -1$, then the cofactors $\prod_{j \neq i} d_j = -1$, and the ith diagonal entry of D^{-1} equals -1. A similar argument for the case where $\det S = -1$ shows that the sign of the ith entries of D and D^{-1} are always the same. Hence $D = D^{-1}$.

Now suppose $B = DAD$ is nonnegative, and let λ be such the $Bx = \lambda x$. Then we have $DB = AD$ because $DD = \mathrm{I}$. It follows that $DBx = \lambda Dx = ADx$, so that λ is also an eigenvalue of A with corresponding eigenvector Dx.

We say that the matrix A is *sign similar* to a nonnegative matrix if $B = SAS$ is a nonnegative matrix for some signature matrix S. The earlier results in this appendix can all be translated so that they apply to such matrices. For example, we have the following analogue of Theorem A5.1.

THEOREM A5.1$'$. *Let A be an irreducible matrix sign similar to a nonnegative matrix. Then there is a maximal eigenvalue λ^* of A that is simple, and for any other eigenvalue λ, $\lambda^* \geq |\lambda|$. If $|\lambda| = \lambda^*$, then λ is also a simple eigenvalue. Moreover, there exists a positive vector x^* such that Sx^* is an eigenvector of A belonging to λ^*, where S is the unique signature matrix such that $B = SAS$ is nonnegative.*

How can matrices sign similar to a nonnegative matrix be recognized? We claim the following result is true.

THEOREM A5.4. *Let A be an irreducible matrix. Then A is sign similar to a nonnegative matrix if and only if there exists a permutation matrix P such that $PAP^T = \tilde{A}$ with*

$$A = \begin{bmatrix} A_{11} & -A_{12} \\ -A_{21} & A_{22} \end{bmatrix},$$

where each $A_{ij} \geq 0$ and A_{11} and A_{22} are square submatrices.

PROOF. Suppose first that such a permutation matrix P exists, and let $S_0 = \text{diag}[1, 1, 1, \ldots, 1, -1, -1, \ldots, -1]$, where the first r entries are 1 if

A_{11} is an $r \times r$ submatrix and the remaining $n-r$ entries are -1. Then $S_0 \tilde{A} S_0 = B$, where B is nonnegative. Hence A is sign similar to a nonnegative matrix. For the converse, suppose S is such that $SAS = B$ is nonnegative, with $S = \text{diag}[d_1, \ldots, d_n]$ and each d_i equal to 1 or -1. Let P_0 be the permutation matrix such that $P_0 S = \text{diag}[1, 1_1, \ldots, 1, -1, \ldots, -1]$, where there are $r > 0$ 1s and $n - r > 0$ $-$ 1s. Then $P_0 A P_0^T$ has the above form.

A5.3 Matrices with Only Positive Cycles

The matrices with only positive cycles are of special interest to us. We prove the following results about such matrices, which we write in the special form $A = A_d + \tilde{A}$; i.e., $A_d = \text{diag}[a_{11}, \ldots, a_{mn}]$, and $\tilde{A} = A - A_d$ is a matrix with zeros on the principal diagonal. We assume in this section that A is irreducible.

LEMMA A.5.1. *Suppose A is a matrix with only positive cycles, and let C_1 be a cycle of A with index set $\{i(1), \ldots, i(k)\}$ $k \geq 2$. Then $\{i(1), \ldots, i(k)\} = T_1 \cup T_2$ (where one of the sets may be empty), $T_1 \cap T_2 = \varnothing$, $a_{ij} \geq 0$ if i and j both belong to either T_1 or T_2, and $a_{ij} \leq 0$ if i belongs to one set and j to the other.*

PROOF. If all of the entries $a_{i(1)i(2)}, \ldots, a_{i(k)i(1)}$ of C_1 are positive, we set $T_1 = \{i(1), \ldots, i(k)\}$ and $T_2 = \varnothing$, and the lemma follows because all nonzero cycles are positive. If an entry $a_{i(j)i(j+1)} < 0$ $(i(k+1) = i(1))$, then we place $i(j)$ and $i(j+1)$ in opposite sets. Because C_1 is positive we uniquely define the sets T_1 and T_2 by placing $i(1)$ in T_1 and proceeding through the cycle C_1. Moreover, we have $T_1 \cap T_2 = \varnothing$. That $a_{ij} > 0$ if it is nonzero and i and j belong to the same set and that $a_{ij} < 0$ if it is nonzero and i and j are in different sets follow from the fact that every nonzero cycle of A is positive.

LEMMA A.5.2. *If A is a matrix with only positive cycles, then the set $N = \{1, 2, \ldots, n\} = T_1 \cup T_2$ where one of the sets may be empty, $T_1 \cap T_2 = \varnothing$, $a_{ij} \geq 0$ if i and j both belong to the same set, and $a_{ij} \leq 0$ if i and j belong to different sets.*

PROOF. Since A is irreducible, every nonzero entry of A belongs to a cycle. We start constructing T_1 and T_2 by choosing the longest cycle C_1 of A. By Lemma A5.1, the principal submatrix containing the indices of C_1 satisfies the conditions of the lemma. If C_1 is a cycle of length n, we are done. If C_1 has length less than n, choose the longest cycle C_2 of A containing indices in C_1 and indices not in C_1. Beginning with an index in C_1, place the new indices in C_2 into T_1 or T_2, following the rules used in the proof of Lemma A5.1. By positivity of cycles, the principal submatrix containing all of the subscripts of C_1 and C_2 satisfies the conditions of the lemma. It is clear that we may continue the process until the lemma is proved because A is irreducible.

THEOREM A5.5. *The matrix A has only positive cycles if and only if there exists a permutation matrix P such that*

$$P\tilde{A}P^T = \begin{bmatrix} \tilde{A}_{11} & \tilde{A}_{12} \\ \tilde{A}_{21} & \tilde{A}_{22} \end{bmatrix},$$

where $\tilde{A}_{ij} \geq 0$ *and* $\tilde{A}_{ii}$ *is a square matrix for* $i = 1, 2$.

PROOF. We can choose P such that T_1 maps into $1, 2, \ldots, r$ and T_2 maps into $r+1, \ldots, n$, where $r \leq n$, and Lemma A5.2 implies the conditions are satisfied. For the converse, assume P exists. Observe that the transformation SBS preserves the signs of the cycles of the matrix B, since if b_{ij} has all signs reversed for fixed j under the transformation, then all b_{jk} also have the signs reversed. Hence in any cycle signs are reserved in pairs. But, by Theorem A5.4, $SP\tilde{A}P^TS$ is sign similar to a nonnegative matrix; hence it has only positive cycles. Thus $P\tilde{A}P^T$ and so also $\tilde{A}$ must have only positive cycles. Thus A has only positive cycles and the theorem is proved.

We now know that A has only positive cycles if and only if $\tilde{A}$ is sign similar to a nonnegative matrix.

A5.4 Metzler Matrices and Morishima Matrices

In economics, traditionally there has been a great deal of interest in matrices related to nonnegative matrices (and to matrices with nonnegative cycles) by the relation $A = M - \alpha I$, where M is a nonnegative matrix (or a matrix with nonnegative cycles) and $\alpha > 0$ satisfies $\alpha > m_{ii}$ $i = 1, \ldots, n$, so that A has diagonal entries negative.

Matrices with negative diagonal and nonnegative off-diagonal entries are known as *Metzler matrices* in economics (and as "gross substitute" matrices). Irreducible matrices with negative diagonal and all nonnegative cycles, but containing some negative off-diagonal entries, are called *Morishima matrices*. The importance of these two classes of matrices to the qualitative analysis of economics systems will become clear in the chapters that follow. Of particular importance is the characterization of stable matrices of these classes. Theorems A5.6 and A5.7 identify certain useful equivalent conditions for stability of Metzler and Morishima matrices. (See Berman and Plemmons [1979] for an exhaustive discussion of these and numerous other equivalent conditions for stability of Metzler and Morishima matrices.)

THEOREM A5.6 (STABLE METZLER MATRICES). *Let* $A = M - \alpha I$, *where* M *is an irreducible* $n \times n$ *matrix with all entries nonnegative and where* $\alpha > m_{ii}$ $i = 1, \dots, n$ *so that* $a_{ii} < 0$ $i = 1, \dots, n$. *Let* λ^* *denote the maximum eigenvalue of* M. *Then* A *is stable if and only if any of the following conditions hold*:

(i) $\alpha > \lambda^*$; *and/or*
(ii) A *is dominant diagonal*; *and/or*
(iii) A *is Hicksian*; *and/or*
(iv) $A^{-1} < 0$; *and/or*
(v) *there exists a positive vector* y *such that* $Ay < 0$.

PROOF. (i) Let p denote an eigenvalue of A such that $\det(A - pI) = 0$ implies $\det(M - (\alpha + p)I) = 0$, so that λ an eigenvalue of M implies $\lambda = \alpha + p$. Since M is an irreducible nonnegative matrix, $|\lambda| \leq \lambda^*$ for all λ implies $\lambda * \geq R(\lambda)$ for all λ. $R(\lambda) = \alpha + R(p)$; hence $R(p) = R(\lambda) - \alpha$. $R(p) < 0$ for all p implies $\lambda^* - \alpha < 0$, or $\alpha > \lambda^*$; and if $\alpha > \lambda^*$, $R(p) < 0$ for all p.

(ii) By (i), A is stable if and only if $\alpha > \lambda^*$. By the Perron-Frobenius theorem, there exists a positive vector x^* such that $Mx^* = \lambda * x^*$, and this is the only eigenvector (up to a scalar multiple) with this property. Hence $|m_{ii} - \lambda^*| x_i^* = \sum_{j \neq i} m_{ij} x_j^*$ $i = 1, \dots, n$. Thus $|a_{ii} + (\alpha - \lambda^*)|$ $x_i^* = \sum_{j \neq i} a_{ij} x_j^* i = 1, \dots, n$. A stable implies $\alpha > \lambda^*$, while $a_{ii} < 0$ so that $|a_{ii}| x_i^* > |a_{ii}(\alpha - \lambda^*)| x_j^* = \sum_{j \neq i} a_{ij} x_j^*$ and A is dominant diagonal. Similarly, if A is dominant diagonal with negative diagonal elements, then it is well known (see, e.g., McKenzie, 1960a) that A is stable regardless of the signs of off-diagonal entries.

(iii) If A is stable, it is dominant diagonal by (ii). Any dominant diagonal matrix has the property that any principal submatrix is dominant diagonal. Since A has all diagonal elements negative, when A is stable, every principal submatrix of A is stable. By the Routh-Hurwitz conditions, this implies that every ith-order principal minor of A has sign $(-1)^i$ $i = 1, \dots, n$, so that A is Hicksian. On the other hand, if A is Hicksian, then A^{-1} exists with all diagonal elements in A^{-1} negative. Since A has all nonzero chains positive, A irreducible implies that all off-diagonal elements in A^{-1} are negative as well. By (iv) (see below) this implies that A is stable.

(iv) Note that since $A = M - \alpha I$, A has the same set of eigenvectors as M. In particular, A possesses a unique (up to a scalar multiple) positive eigenvector x^* such that $Ax^* = \rho^* x^*$, where $\rho^* = (\lambda^* - \alpha)$. Suppose A^{-1} exists with $A^{-1} < 0$. Then $x^* = (\lambda^* - \alpha)A^{-1}x^*$, which in turn implies $\lambda^* - \alpha < 0$; i.e., A is stable.

Similarly, suppose A is stable so that A^{-1} exists. By (iii) A stable implies A is Hicksian, which together with the fact that all nonzero chains in A are positive implies $A^{-1} < 0$.

(v) In $Ay = \rho y$, we have seen that $Ay^* = \rho^* y^*$, where $\rho^* = (\lambda^* - \alpha)$ and $y^* = x^*$, a strictly positive vector. Thus $Ax^* = (\lambda^* - \alpha)x^*$. If $\alpha > \lambda^*$, then A is stable, which implies that x^* is a positive vector for which $Ax^* < 0$. On the other hand, if $Ay < 0$ for some positive vector y, then $|a_{ii}|y_i > \sum_{j \neq i} a_{ij} y_j$ $i = 1, \ldots, n$, and A is dominant diagonal and hence stable.

As noted earlier, a Morishima matrix B is similar to a Metzler matrix, $B = SAS^{-1}$ where A is a Metzler matrix and S is a diagonal matrix with $s_{ii} > 0$ $i = 1, \ldots, r$, $s_{ii} < 0$ $i = r+1, \ldots, n$ (or conversely). An irreducible Metzler matrix can be written in block form as

$$B = \begin{bmatrix} B_1 & B_2 \\ B_3 & B_4 \end{bmatrix},$$

where B_1 and B_4 are square Metzler matrices and $B_2 \leq 0$, $B_3 \leq 0$. We here state without proof the equivalent conditions for stability of Morishima matrices, corresponding to those of the previous theorem.

THEOREM A5.6 (STABILITY OF MORISHIMA MATRICES). *Let $B = SAS^{-1}$ be an $n \times n$ irreducible Morishima matrix, where $A = M - \alpha I$, M is an irreducible nonnegative matrix. $B = \begin{bmatrix} B_1 & B_2 \\ B_3 & B_4 \end{bmatrix}$, where B_1 and B_4 are square Metzler submatrices and $B_2 \leq 0$, $B_3 \leq 0$. Then B is stable if and only if any of the following hold:*

(i) $\alpha > \lambda^*$; *and/or*
(ii) *B is dominant diagonal; and/or*
(iii) *B is Hicksian; and/or*
(iv) *letting $C = B^{-1}$, where $C = \begin{bmatrix} C_1 & C_2 \\ C_3 & C_4 \end{bmatrix}$, partitioned as in B, $C_1 < 0$, $C_4 < 0$, and $C_2 > 0$, $C_3 > 0$; and/or*
(v) *there exists a vector $y = (y^1, y^2)$, partitioned as B, with $y^1 > 0$, $y^2 < 0$ such that $By < 0$.*

6

The Correspondence Principle

6.1 INTRODUCTION

This chapter deals with Samuelson's (1947) "correspondence principle" as it applies to qualitative models. The underlying idea is this: Comparative statics analysis in economics and in other fields is concerned with deriving predictive statements about changes in the values of certain endogenous variables in response to changes in the values of certain exogenous variables. For example, economists attempt to predict the change in price of a good that will occur if an excise tax is imposed on it, or the change in Gross Domestic Product (GDP) that will occur if the Federal Reserve increases the money supply. The conceptual apparatus employed is one in which it is assumed that the initial state of the model is an equilibrium state and that the prediction that is made relates to the changes in the endogenous variables at "before" and "after" equilibrium states of the model in response to postulated changes in the exogenous valuables.

Under the correspondence principle, it is assumed that the initial ("before") equilibrium state that is observed is a stable equilibrium, i.e., if the equilibrium were disturbed, the dynamic adjustment mechanism would act to reestablish equilibrium. Samuelson argues convincingly that the chances of actually observing an unstable equilibrium are comparable to those of observing an egg standing on end, since at an unstable equilibrium, the slightest disturbance of the system causes the system to diverge from its initial equilibrium. Figure 6.1 illustrates this for the case of an isolated market, where the dynamics of the system follow the usual rules: price rises when excess demand is positive, and price falls when excess demand is negative.

In Figure 6.1, p^* and p^{***} are stable equilibrium positions, as indicated by the arrows pointing in the direction of change in price, given the dynamic adjustment mechanism; p^{**} is an unstable equilibrium. Under the correspondence principle, it is assumed that observed equilibrium positions are such as p^* and p^{***}, rather than p^{**}. What Samuelson argues is that the (reasonable) assumption that the observed equilibrium is a dynamically stable equilibrium leads, in certain cases at

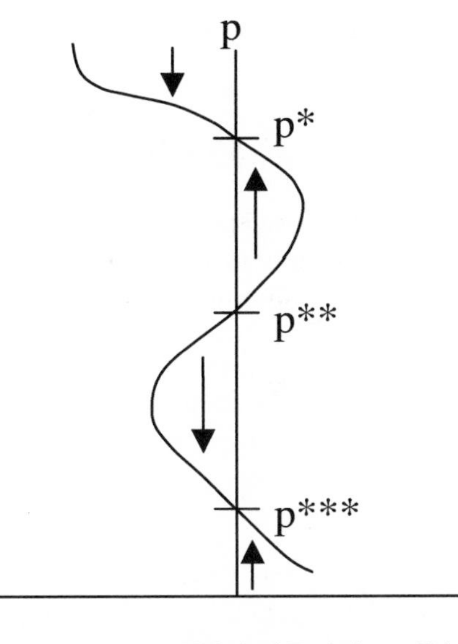

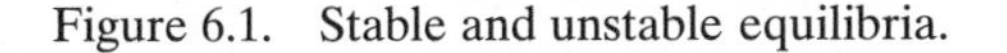

Figure 6.1. Stable and unstable equilibria.

least, to restrictive comparative statics conclusions. That is, corresponding to stable equilibrium positions are comparative statics results concerning the effects on equilibrium values of disturbances to equilibrium.

In the context of qualitatively specified systems, the correspondence principle can be formulated as follows. Given the qualitative matrix $Q_A = \{B \mid \operatorname{sgn} B = \operatorname{sgn} A\}$, consider the subset of Q_A (perhaps empty), denoted by Q_A^*, which consists of those matrices in Q_A that are stable matrices, i.e., $B \in Q_A^*$ implies the real parts of the eigenvalues of B are negative. The correspondence principle then argues that, in certain cases at least, restricting attention to Q_A^* rather than Q_A, can lead to predictive comparative statics statements, here interpreted as full or partial sign solvability of the system $Ax = b$. However, Samuelson gives only a few examples of applications of the correspondence principle, and it remains an open question as to the classes of cases in which the stability hypothesis (together with other structural information, such as qualitative information) can lead to restrictive comparative statics theorems. An early discussion of this issue appears in Patinkin (1952);

see also Bassett, Maybee, and Quirk 1968. This chapter provides a partial identification of the class of qualitatively specified cases in which the correspondence principle in fact applies.

6.2 AN EXAMPLE

To illustrate, consider the correspondence principle applied to the comparative statics of a two-market partial equilibrium system. Excess demand for goods 1 and 2 are denoted by E_1 and E_2, where

$$E_1 = E_1(p_1, p_2, \alpha),$$

and

$$E_2 = E_2(p_1, p_2).$$

In these expressions, p_1 and p_2 are the prices of good 1 and good 2, and α is a shift parameter, e.g., the level of a tax that affects the production costs of good 1. For a given value of α, equilibrium of this system of markets occurs at a price vector $p^*(\alpha) = (p_1^*(\alpha), p_2^*(\alpha))$ such that

$$E_1(p_1^*(\alpha), p_2^*(\alpha), \alpha) = 0 \text{ and } E_2(p_1^*(\alpha), p_2^*(\alpha)) = 0.$$

The comparative statics analysis of this system investigates the effect on p^* of a change in α. The change in equilibrium prices following a change in α is given by

$$\frac{dE_1}{d\alpha} = \frac{\partial E_1}{\partial p_1}\frac{dp_1^*}{d\alpha} + \frac{\partial E}{\partial p_2}\frac{dp_2^*}{d\alpha} + \frac{\partial E_1}{\partial \alpha} = 0,$$

and

$$\frac{dE_2}{d\alpha} = \frac{\partial E_2}{\partial p_1}\frac{dp_1^*}{d\alpha} + \frac{\partial E_2}{\partial p_2}\frac{dp_2^*}{d\alpha} = 0.$$

The comparative statics system we wish to study is $Ax = b$, where

$$A = \begin{bmatrix} \dfrac{\partial E_1}{\partial p_1} & \dfrac{\partial E_1}{\partial p_2} \\ \dfrac{\partial E_2}{\partial p_1} & \dfrac{\partial E_2}{\partial p_2} \end{bmatrix}, x = \begin{bmatrix} \dfrac{dp_1^*}{d\alpha} \\ \dfrac{dp_2^*}{d\alpha} \end{bmatrix} b = -\begin{bmatrix} \dfrac{\partial E_1}{\partial \alpha} \\ 0 \end{bmatrix}.$$

Suppose the dynamic adjustment of prices takes the form

$$\dot{p}_1 = \beta_1 E_1(p_1, p_2, \alpha)$$
$$\dot{p}_2 = \beta_2 E_2(p_1, p_2) \qquad (\beta > 0, \beta_2 > 0).$$

In linear approximation form, this becomes

$$\dot{p}_1 = \beta_1 \left\{ \frac{\partial E_1}{\partial p_1}(p_1 - p_1^*) + \frac{\partial E_1}{\partial p_2}(p_2 - p_2^*) \right\}$$
$$\dot{p}_2 = \beta_2 \left\{ \frac{\partial E_2}{\partial p_1}(p_1 - p_1^*) + \frac{\partial E_2}{\partial p_2}(p_2 - p_2^*) \right\}.$$

Again, in matrix notation, this becomes

$$\dot{p} = BA(p - p^*),$$

where $B = \begin{bmatrix} \beta_1 & 0 \\ 0 & \beta_2 \end{bmatrix}$, $\beta_1 > 0, \beta_2 > 0$.Assume the law of demand holds so that $\dfrac{\partial E_i}{\partial p_i} < 0 \ i = 1,2$. We look at the following special cases:

(i) $$A = \begin{bmatrix} - & + \\ + & - \end{bmatrix}$$

(ii) $$A = \begin{bmatrix} - & + \\ - & - \end{bmatrix}$$

(iii) $$A = \begin{bmatrix} - & - \\ + & - \end{bmatrix}$$

(iv) $$A = \begin{bmatrix} - & - \\ - & - \end{bmatrix}$$

Note that for a given α, along the curve $E_1 = 0$, we have

$$\left.\frac{dp_2}{dp_1}\right|_{E_1=0} = (-)\frac{\partial E_1}{\partial p_1} \Big/ \frac{\partial E_1}{\partial p_2},$$

and similarly, along $E_2 = 0$, we have

$$\left.\frac{dp_2}{dp_1}\right|_{E_2=0} = (-)\frac{\partial E_2}{\partial p_1} \bigg/ \frac{\partial E_2}{\partial p_2}.$$

Phase diagrams for the four cases under the stability hypothesis are shown in Figure 6.2. In cases (ii) and (iii), the system is sign stable, as discussed in Chapter 2, so the stability hypothesis adds no new restrictions to the analysis. In cases (i) and (iv), stability occurs only if the determinantal condition $a_{11}a_{22} - a_{12}a_{21} > 0$ is satisfied. This determines the relative slopes of the $E_1 = 0$ and $E_2 = 0$ curves in (i) and (iv). The fact that the phase diagrams all exhibit stability is indicated by the arrows in all quadrants pointing toward the equilibrium, indicating the direction of the time paths for p_1 and p_2 when they enter those quadrants, from the differential equation dynamic adjustment mechanism.

Suppose now that there is an increase in α that acts to shift the $E_1 = 0$ curve to the right. The effect on the equilibrium values of p_1 and p_2 are shown in Figure 6.3. In (i) and (iii), a shift in E_1 to the right increases p_1^* and p_2^*, while in (ii), the shift increases p_1^* and reduces p_2^*. Further, because each of the situations exhibits dynamic stability, this means that in each case, prices move over time to converge to the new equilibrium; i.e., the comparative statics theorems have predictive power.

The algebraic solutions to the comparative statics problems are as follows.

$$\text{Let } x_1 = \frac{dp_1^*}{d\alpha}, x_2 = \frac{dp_2^*}{d\alpha}.$$

(i)

$$\begin{bmatrix} - & + \\ + & - \end{bmatrix}\begin{bmatrix} x_1 \\ x_2 \end{bmatrix} = \begin{bmatrix} - \\ 0 \end{bmatrix} \Rightarrow \begin{bmatrix} x_1 \\ x_2 \end{bmatrix} = \begin{bmatrix} - & - \\ - & - \end{bmatrix}\begin{bmatrix} - \\ 0 \end{bmatrix} \Rightarrow \begin{bmatrix} x_1 \\ x_2 \end{bmatrix} = \begin{bmatrix} + \\ + \end{bmatrix}$$

(ii)

$$\begin{bmatrix} - & + \\ - & - \end{bmatrix}\begin{bmatrix} x_1 \\ x_2 \end{bmatrix} = \begin{bmatrix} - \\ 0 \end{bmatrix} \Rightarrow \begin{bmatrix} x_1 \\ x_2 \end{bmatrix} = \begin{bmatrix} - & - \\ + & - \end{bmatrix}\begin{bmatrix} - \\ 0 \end{bmatrix} \Rightarrow \begin{bmatrix} x_1 \\ x_2 \end{bmatrix} = \begin{bmatrix} + \\ - \end{bmatrix}$$

(iii)

$$\begin{bmatrix} - & - \\ + & - \end{bmatrix}\begin{bmatrix} x_1 \\ x_2 \end{bmatrix} = \begin{bmatrix} - \\ 0 \end{bmatrix} \Rightarrow \begin{bmatrix} x_1 \\ x_2 \end{bmatrix} = \begin{bmatrix} - & + \\ - & - \end{bmatrix}\begin{bmatrix} - \\ 0 \end{bmatrix} \Rightarrow \begin{bmatrix} x_1 \\ x_2 \end{bmatrix} = \begin{bmatrix} + \\ + \end{bmatrix}$$

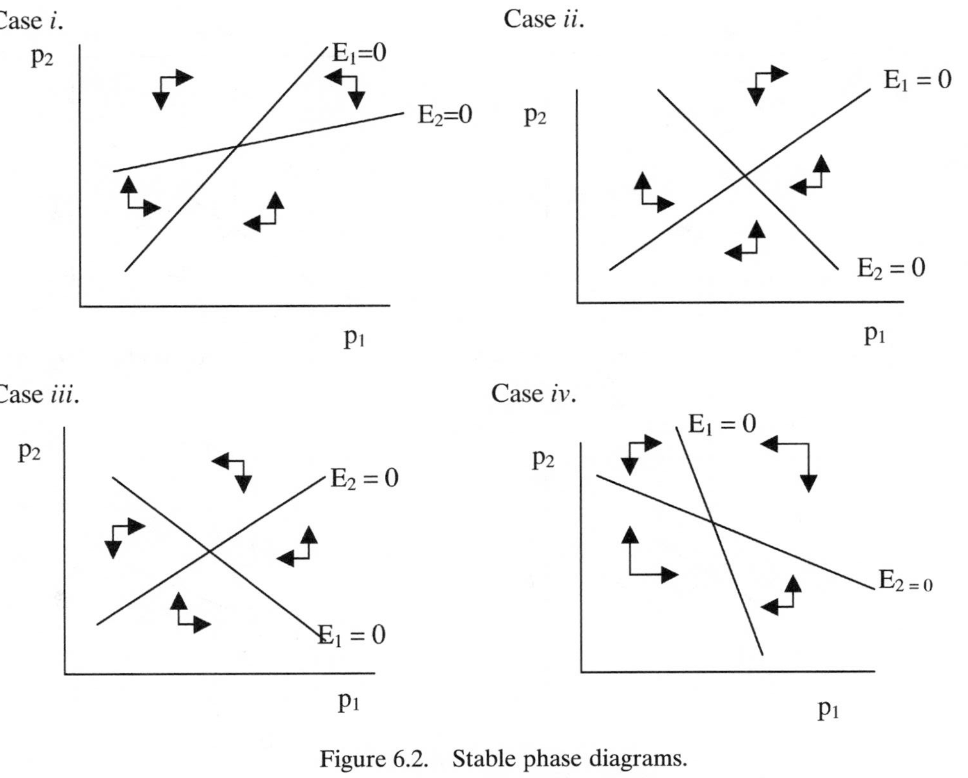

Figure 6.2. Stable phase diagrams.

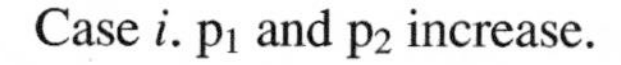
Case *i*. p_1 and p_2 increase.

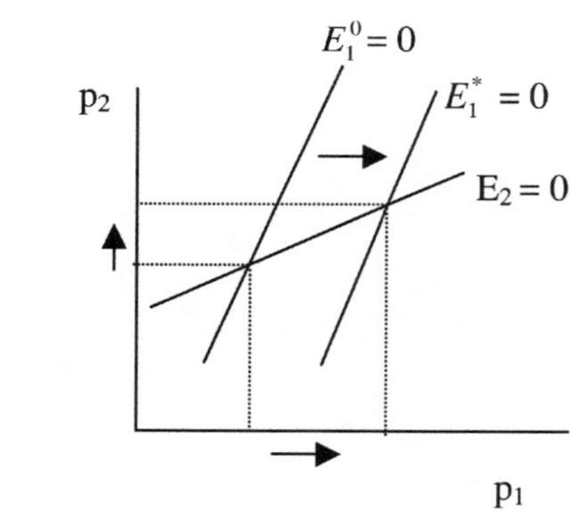

Case *ii*. p_1 increases and p_2 decreases.

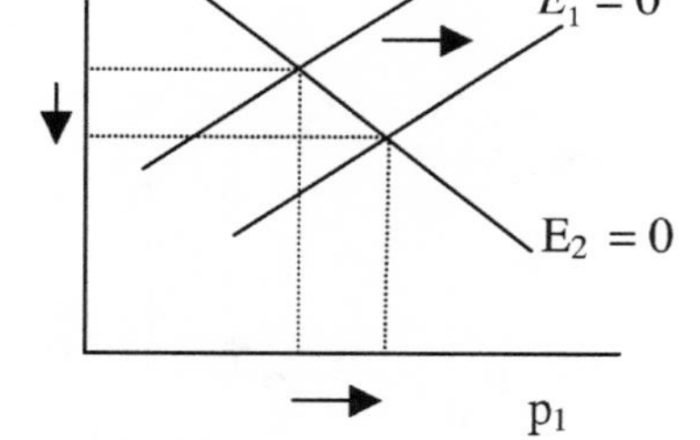
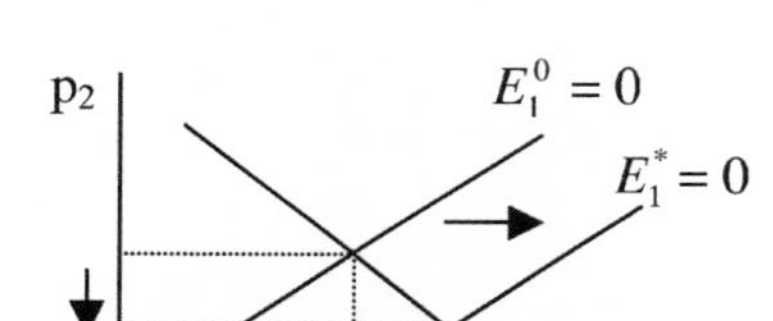

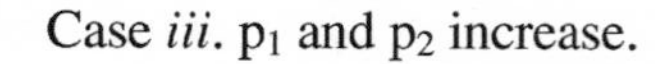
Case *iii*. p_1 and p_2 increase.

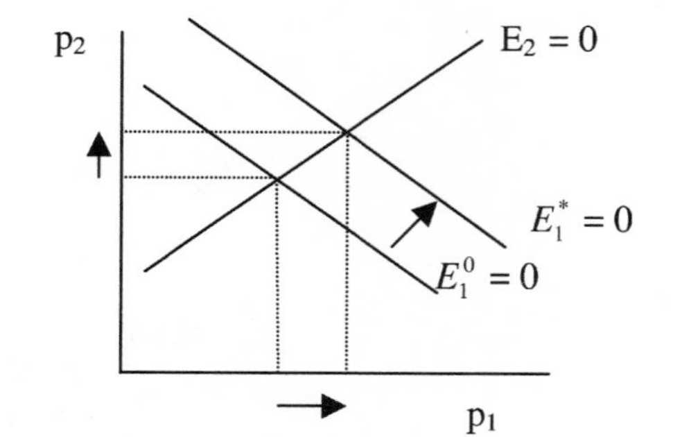

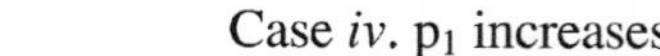
Case *iv*. p_1 increases and p_2 decreases.

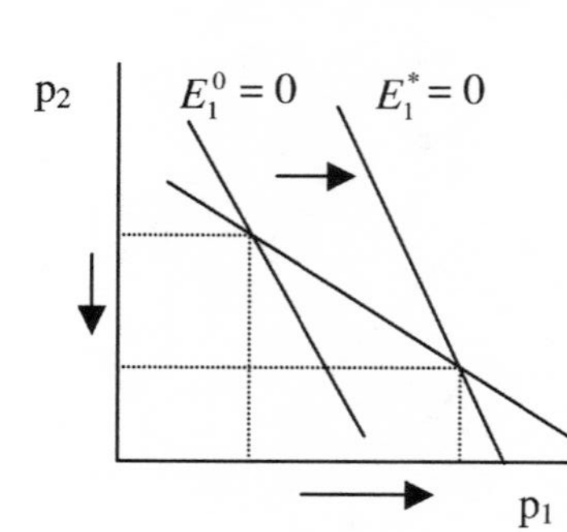
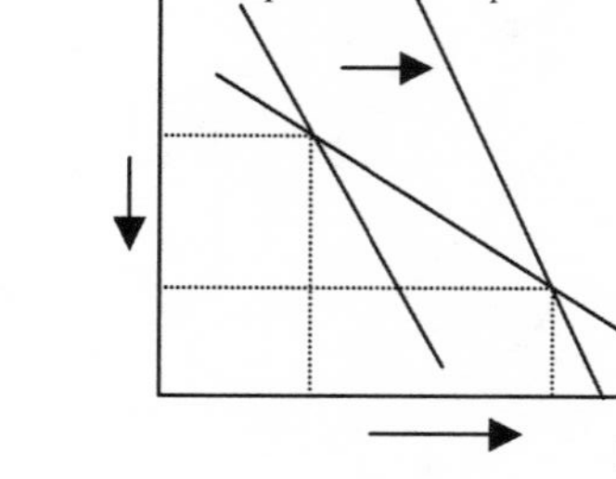

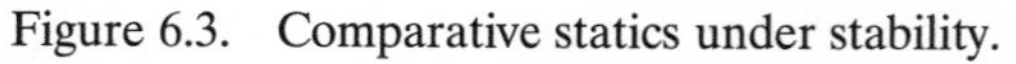
Figure 6.3. Comparative statics under stability.

(iv)

$$\begin{bmatrix} - & - \\ - & - \end{bmatrix}\begin{bmatrix} x_1 \\ x_2 \end{bmatrix} = \begin{bmatrix} - \\ 0 \end{bmatrix} \Rightarrow \begin{bmatrix} x_1 \\ x_2 \end{bmatrix} = \begin{bmatrix} - & + \\ + & - \end{bmatrix}\begin{bmatrix} - \\ 0 \end{bmatrix} \Rightarrow \begin{bmatrix} x_1 \\ x_2 \end{bmatrix} = \begin{bmatrix} + \\ - \end{bmatrix}$$

Note that in (i) and (iv), the stability hypothesis is involved in that the assumption that equilibrium is stable leads to a signing of the determinant of the coefficient matrix as positive, which in turn leads to the comparative statics results indicated.

6.3 STABILITY AND SIGN SOLVABILITY

This section summarizes results linking stability to sign solvability. As noted earlier, an $n \times n$ matrix A is said to be a stable matrix if the real parts of all eigenvalues of A are negative. Let $Q_A^* = \{B \mid B \in Q_A$ and B is a stable matrix$\}$. $Ax = b$ is said to be sign solvable under the stability hypothesis if $B \in Q_A^*$, $c \in Q_b$, and $By = c$ implies $y \in Q_x$. As in the earlier cases, $Ax = b$ is partially sign solvable under the stability hypothesis if there exists a subset of the entries of x that is sign solvable under that hypothesis.

One of the fundamental problems with determining the scope of the correspondence principle is determining the class of qualitatively specified matrices A for which Q_A^* is nonempty. If there exists $B \in Q_A$ such that Q_A^* is nonempty, then A is said to be a *potentially stable* matrix. There is a limited amount of work on the potential stability problem. We have discussed what is known in Section 2.6 of Chapter 2.

Clearly, if A has all diagonal entries negative, then A is a potentially stable matrix. Essentially all the work that has been done on sign solvability under the stability hypothesis has dealt only with this case, and, within the class of negative diagonal matrices, results on sign solvability exist only for the case where dynamic stability of A implies Hicksian stability of A, i.e., where every ith-order principal minor of A has sign $(-1)^1$, $i = 1, \ldots, n$.

The notion of Hicksian stability entered the economics literature following the publication of Hicks's seminal work, *Value and Capital*, in 1939. Hicks reintroduced general equilibrium theory into the mainstream of economics, building on early work by Walras. In his discussion of the stability of a general equilibrium of the economy, Hicks attempted to generalize to the n-dimension case the notion of stability for a single market. His heuristic approach led him to the idea that there would be stability in a system of markets if for each commodity, excess demand for the commodity, which in general depends on the prices of all commodities in the system, is a decreasing function of the price of the commodity, given that any arbitrary set of prices for other commodities can be adjusted, so long as the markets for those commodities whose prices have adjusted are cleared. Hicks referred to this as "perfect stability" of markets, and the literature since *Value and Capital* has generally referred to it as "Hicksian stability." As shown in the discussion of Hicksian stability in the appendix to Chapter 8 (see also Quirk and Saposnik 1968), Hicksian stability in the neighborhood of equilibrium occurs if and only if the matrix A of partial derivatives of excess demand with respect to prices obeys the rule that ith-order principal minors have sign $(-1)^i$ for $i = 1, \ldots, n$.

There was a lively dispute in economics over Hicks's notion of stability following publication of *Value and Capital* Samuelson (1944) correctly pointed out that Hicks had introduced a static notion of stability to solve a problem that was inherently dynamic in nature. Samuelson argued that the correct generalization of single-market stability was true dynamic stability, as applied in the physical sciences. He also pointed out that the appropriate dynamic adjustment process, again referring back to Walras, was the *tâtonnement* process, under which the price of a commodity is increased if excess demand for the commodity is positive and price is decreased if excess demand for the commodity is negative. Stability of this system, then, is asymptotic stability, asymptotic convergence of the set of prices to an equilibrium for arbitrary initial conditions.

This led to further research concerning the links between Samuelson's dynamic stability and Hicksian stability. Metzler (1945) provided examples to show that in the general case, Hicksian stability is neither necessary nor sufficient for dynamic stability. On the other hand, Metzler showed that if there is dynamic stability of every subset of markets for all (positive) speeds of adjustment of markets (this has since been termed *total stability*), then this implied Hicksian stability. As noted earlier, the Fuller and Fisher (1958) result asserts, among other things, that if A is a Hicksian matrix, then there exists a diagonal matrix D with all diagonal elements positive such that DA has all eigenvalues with negative real parts, i.e., such that DA is a stable matrix. Also, as indicated in the appendix to Chapter 5, in the special cases of Metzler or Morishima matrices, A is stable if and only if it is Hicksian stable.

Samuelson's approach to modeling stability under a *tâtonnement* process is, of course, the correct concept to apply to economic models to describe their behavior while they adjust over time to disturbances to equilibrium under a *tâtonnement* mechanism. On the other hand, for comparative statics analysis, Hicks's notion of stability is the useful concept, since it provides immediate information on important characteristics of A^{-1}. For this reason, links between dynamic and Hicksian stability are of particular interest in comparative statics analysis.

Consider first two polar cases of matrices with negative diagonals that have the property that dynamic stability of A implies Hicksian stability of A: namely, the case in which all nonzero cycles are negative and the case where all nonzero cycles are positive. In the first case, the case of an L-matrix or a sign nonsingular (SNS) matrix (see the appendix to chapter 2), any such matrix is Hicksian simply on the basis of its sign pattern alone. In the second case, stability is equivalent to the Hicksian property; if the matrix is also irreducible it is either a Metzler or a Morishima matrix. In both cases, derivation of necessary and sufficient conditions for sign solvability under the stability hypothesis is straightforward (Bassett, Maybee, and Quirk 1968).

THEOREM 6.1. *Assume A has all diagonal entries negative, with all cycles in A nonpositive. Then $Ax = b$ is sign solvable under the stability hypothesis if and only if $Ax = b$ is sign solvable in the purely qualitative case; i.e., $b_k \neq 0$ implies that all nonzero paths $a(i \to k)$ have the same sign for $i \neq k$, $i, k = 1, \ldots, n$, with the signs of the paths such that all terms $b_k A_{ki}$ have weakly the same sign.*

PROOF. The condition on the paths is sufficient for sign solvability in the purely qualitative case, given that all nonzero cycles in A are negative, and hence is sufficient for sign solvability under the stability hypothesis. To show that sign solvability under stability implies sign solvability in the purely qualitative case reduces to showing that the path condition is necessary as well as sufficient for sign solvability under the stability hypothesis.

Assume without loss of generality that $b_1 \neq 0$ and there exist two nonzero paths $a^*(2 \to 1)$ and $a^{**}(2 \to 1)$ of opposite signs. Choose $B \in Q_A$ with all diagonal entries $b_{ii} = -2$, and assign each entry in the path $a^*(2 \to 1)$ an absolute value of 1, with all other nonzero entries in B having an absolute value of ε. Choose $C \in Q_A$ with the same assignment of values except that now it is the entries in $a^{**}(2 \to 1)$ that are assigned an absolute value of 1, with all other nonzero off-diagonal entries in C being assigned a value of ε. As ε approaches zero, consider the naturally ordered nested sequence of principal minors of B (or C), beginning with the diagonal entry in the first row and column and ending with the determinant of the matrix. In both cases, the nested sequence of principal minors satisfies the condition that the ith-order principal minor in the sequence has sign $(-1)^i$ $i = 1, \ldots, n$. Thus, by the Fisher-Fuller theorem, there exists a diagonal matrix D with positive diagonal elements such that DB is a stable matrix (and similarly, there exists such a diagonal matrix E so that EC is stable). $DB \in Q_A^*$ (and $EC \in Q_A^*$). However, as ε approaches zero, sgn $DB_{12} = -\text{sgn } a^*(2 \to 1)$, while sgn $DC_{12} = -\text{sgn } a^{**}(2 \to 1)$. Therefore, sgn $DB_{12} \neq \text{sgn } EC_{12}$. Hence $Ax = b$ is not sign solvable under the stability hypothesis.

The argument underlying the proof of Theorem 6.1, leads to the following results as well.

COROLLARY 1. *Under the conditions of Theorem* 6.1, $Ax = b$ *is partially sign solvable under the stability hypothesis if and only if it is partially sign solvable in the purely qualitative case.*

COROLLARY 2. *Let* A *be a matrix satisfying the condition that* $B \in Q_A^*$ *implies that* B *is a Hicksian matrix. Then a necessary and sufficient condition for* $Ax = b$ *to be fully sign solvable under the stability hypothesis is the path condition given in Theorem* 6.1.

THEOREM 6.2. *Given that* A *has all diagonal elements negative, a necessary condition for* $Ax = b$ *to be fully sign solvable under the stability hypothesis is the path condition given in Theorem* 6.1.

PROOF. If all ith-order principal minors are signed $(-1)^i$, $i = 1, \ldots, n$, then the theorem follows from Corollary 2 to Theorem 6.1. The theorem holds even when there are unsigned principal minors under the stability hypothesis. If the path condition relating to signing the cofactor A_{ij}, $i \neq j$, is violated, either the cofactor is unsigned because of the unsigned principal minor, or it is unsigned because products of signed principal minors with paths enter with different signs in the expansion of A_{ij}. The latter can be shown by using the argument given in the proof of Theorem 6.1. Note that if A has all diagonal entries negative, then there cannot exist ith-order principal minors that are signed with sign $(-1)^{i+1}$ or 0 under the stability hypothesis.

The case of nonnegative cycles is the standard Metzler (or Morishima) case.

THEOREM 6.3. *Let* A *be an irreducible negative diagonal matrix with all nonzero cycles positive; i.e.,* A *is a Metzler matrix or a Morishima matrix. Then* $Ax = b$ *is sign solvable under the stability hypothesis in the Metzler case if and only if every nonzero entry in* b *has the same sign; and* $Ax = b$ *is partially sign solvable under the stability hypothesis if and only if it is sign solvable under that hypothesis.*

In the Morishima case, let the index set of one diagonal Metzler submatrix of A *be* I_1 *and let the index set of the other submatrix be* I_2*. Then* $Ax = b$ *is sign solvable under the stability hypothesis if and only if all* $b_j \neq 0$ *with indices in* I_1 *are of the same sign opposite to that of the*

$b_j \neq 0$ with indices in I_2; and $Ax = b$ is partially sign solvable under the same conditions.

PROOF. Immediate by noting that under stability, if A is a Metzler matrix, then A^{-1} has all entries negative. If A is a Morishima matrix and stable, A^{-1} has negative entries in the Metzler diagonal submatrix and positive entries in the off-diagonal submatrix.

6.4 GM-MATRICES

As it turns out, there is a more general class of matrices, where qualitative stability and the Hicksian property are closely related. This class, first identified by Quirk (1974; see also Maybee and Richman 1988), includes matrices with both positive and negative cycles, as well as L-matrices and Metzler (and Morishima) matrices. Typical elements of this class are referred to as generalized Metzlerian matrices (GM-matrices), defined as follows.

Let A be a matrix with negative diagonal elements. Let I denote the index set of a negative cycle in A, and let J denote the index set of a positive cycle in A. A is said to be a *GM-matrix* if $I \cap J \neq \varnothing$ implies $I \subseteq J$. Examples of GM-matrices include the following:

$$\text{(i)} \quad \begin{bmatrix} - & + & + \\ + & - & + \\ + & + & - \end{bmatrix}$$

$$\text{(ii)} \quad \begin{bmatrix} - & + & - \\ + & - & - \\ - & - & - \end{bmatrix}$$

$$\text{(iii)} \quad \begin{bmatrix} - & + & 0 \\ - & - & + \\ 0 & - & - \end{bmatrix}$$

$$\text{(iv)} \quad \begin{bmatrix} - & + & 0 \\ - & - & + \\ + & 0 & - \end{bmatrix}$$

Since (i) and (ii) (a Metzler and a Morishima matrix, respectively) contain no negative cycles and (iii) (a sign stable matrix) contains no positive cycles, trivially those are members of the GM class. In (iv), the only positive cycle is $a_{12}a_{23}a_{31}$; thus (iv) is GM as well. In contrast, the matrix

$$\begin{bmatrix} - & + & + \\ - & - & + \\ + & - & - \end{bmatrix}$$

is not GM, since $a_{12}a_{21} < 0$ and $a_{13}a_{31} > 0$ have a common index 1 but the index 2 is not in the positive cycle. One important property of GM-matrices is noted in the next theorem.

THEOREM 6.4. *Assume that A is an $n \times n$ matrix with all diagonal entries negative. If A does not belong to the GM class, then there exists $B \in Q_A^*$ such that B is not Hicksian; i.e., there exists an unsigned principal minor of A under the stability hypothesis.*

PROOF. Note first that if the only positive cycles in A are of length n, then $B \in Q_A^*$ implies that B is Hicksian, since the stability hypothesis implies that $\operatorname{sgn}\det B = (-1)^n$, regardless of the cyclic structure of B, while all principal minors of order less than n contain only nonpositive cycles and hence have the Hicksian sign pattern. Suppose then that there exists a positive cycle in A of length less than n, and with index set J. Let I denote the index set of a negative cycle in A such that $I \cap J \neq \varnothing$ and $I \not\subseteq J$. Reindex the positive cycle into $a_{r,r+1}, \ldots, a_{n-1,n}a_{nr}$. Let t denote an index common to I and J, and let s denote an index in I that is not in J. Choose $B \in Q_A$ with $b_{ii} = -1$, $i = 1, \ldots, n$, and let all other nonzero entries in B be assigned an absolute value of ε. For ε sufficiently small, the naturally ordered nested sequence of principal minors obeys the conditions of the Fuller-Fisher theorem, so there exists a diagonal matrix D with diagonal entries positive such that $DB \in Q_A$, DB is stable, and all ith-order principal minors have sign $(-1)^i$, $i = 1, \ldots, n$.

Choose $C \in Q_A$ with $|c_{ij}| = 2$ if c_{ij} belongs to the negative cycle with index set I or if it belongs to the positive cycle with index set J. Let $c_{ii} = -1$ if $i \neq s$, let $c_{ss} = -\varepsilon$, and let all other nonzero entries in C be assigned an absolute value of ε. Once again, consider the naturally ordered nested sequence of principal minors of C. For ε sufficiently small, all elements in that sequence from 1 through $n-1$ obey the condition that ith-order principal minors have sign $(-1)^i$, since products of diagonal terms and/or terms involving the product of the negative cycle with index set I and diagonal entries dominate all other terms in such principal minors. The positive cycle with index set J shows up only in the nth principal minor in the sequence of nested principal minors, but because it is multiplied by a principal minor that is dominated by the product of diagonal terms including $c_{ss}(= -\varepsilon)$, the sign of the nth principal minor is determined by the product of the negative cycle with index set I and its complementary principal minor. Hence again the Fuller-Fisher theorem applies, and a stable matrix can be constructed from the product of a diagonal matrix with positive diagonal elements and C; hence this matrix belongs to $B \in Q_A^*$. However, the $(n-1)$th-order principal minor C_{ss} has sign determined by the product of the positive cycle with diagonal elements; hence $\operatorname{sgn} C_{ss} = (-1)^n$, so that A_{ss} is not signed under the maximization hypothesis.

The GM conditions turn out to be sufficient as well as necessary for stability to imply the Hicksian property for qualitatively specified matrices, as indicated in Theorem 6.5.

THEOREM 6.5. *Assume A is an $n \times n$ matrix that satisfies the GM conditions. Then $B \in Q_A^*$ implies that B is Hicksian.*

PROOF. The proof is by contradiction. Given any GM-matrix A containing negative and positive cycles, suppose that there is an unsigned principal minor in A under the stability hypothesis. As noted earlier, for a principal minor to be unsigned means that there exist $B, C \in Q_A^*$ such that the principal minor is negative in one matrix and positive in the other, with B and C both being stable matrices with the same sign pattern as A.

From the earlier theorems, a GM-matrix containing only nonpositive cycles or only nonnegative cycles has the property that $B \in Q_A^*$ implies that B is Hicksian. Given A, consider a matrix G, say, with the same sign pattern as A except that elements in A appearing in positive cycles only are set equal to zero and the entries in G are chosen so that G is a stable matrix. Of course, G is Hicksian simply because it contains only nonpositive cycles. By continuity, if sufficiently small values are assigned to those elements appearing only in positive cycles in A, this will transform G into a stable matrix H with the same sign pattern as A, and obeying the Hicksian property. Now consider any path through Q_A^* beginning at H, and formed by continuous variations in the values assigned to the nonzero elements of the matrix, while preserving its sign pattern.

Suppose that there is an unsigned principal minor under the stability hypothesis. Then for some initial stable matrix H and for some path through Q_A^*, by continuity, there will exist $B \in Q_A^*$ such that some principal minors(s) in B have value zero, while all remaining principal minors obey the Hicksian property. B can be thought of as a transition matrix from the subset of Q_A^* that contains Hicksian matrices to the subset that contains non-Hicksian matrices. Since ambiguously signed principal minors take on both negative and positive values within Q_A^*, such a B must exist.

For any principal minor in B with value zero, consider the smallest subprincipal minor within it, which is zero. Such a minimal subprincipal minor with index set I will have a positive cycle with index set I. (It must have a positive cycle to take on the value zero, and if the longest positive cycle has an index set strictly contained in I, we have not identified a minimal subprincipal minor). Let $B(I)$ denote the minimal subprincipal minor, and let $B(I')$ denote the principal minor with index set as the complement of I in $\{1,\ldots,n\}$.

The determinant of B can be written as $\det B = B(I)B(I') +$ sum of products of principal minors of B with cycles whose index sets contain elements from both I and I'. By hypothesis, $B(I) = 0$; and, by the GM conditions, since $B(I)$ contains a positive cycle with index set I, all cycles with indices from both I and I' must be nonnegative. But by the Maybee determinantal formula (see the appendix to Chapter 2), the product of a positive cycle with a principal minor obeying the Hicksian sign rules always enters into the expansion of $\det B$ with incorrect sign, i.e., with sign $(-1)^{n+1}$. Thus, $\operatorname{sgn}\det B = (-1)^{n+1}$. However, by the Routh-Hurwitz conditions, B a stable matrix implies that $\operatorname{sgn}\det B = (-1)^n$. This contradiction establishes the theorem.

Theorems 6.4 and 6.5 identify the class of qualitatively specified matrices for which the stability hypothesis "signs" the diagonal entries of the inverse matrix (all must be negative). The question arises as to the implications of the GM conditions for off-diagonal entries of the inverse matrix, where such entries are signed if and only if they obey the path condition of Theorem 6.1. One result is the following.

LEMMA 6.1. *Let A be a GM-matrix, let I denote the index set of a negative cycle in A, and let J denote the index set of a positive cycle in A such that $I \cap J \neq \varnothing$. Then, for every entry a_{ij} appearing in the negative cycle, either A_{ij} is unsigned under the stability hypothesis or A_{ji} is unsigned.*

PROOF. Let $a_{ij}a(j \to i)$ denote a negative cycle with index set I, and let $a_{sr}a^*(r \to s)$ denote a positive cycle with index set J such that at least one index of I appears in J. Consider the cofactor A_{ij}. By the GM conditions, $I \cap J \neq \varnothing$ implies that both i and j appear in the positive cycle J. Since i and j appear in the positive cycle, that cycle can be written as $a^*(i \to j)a^*(j \to i)$. Both $a(j \to i)$ and $a^*(j \to i)$ appear in A_{ij}. Suppose that both have the same sign, so that the path condition is satisfied for A_{ij}. Consider now the cofactor A_{ji}. Paths appearing in A_{ji} include a_{ij} and $a^*(i \to j)$. Since $a_{ij}a(j \to i)$ and $a^*(i \to j)a^*(j \to i)$ have opposite signs, A_{ji} has two paths of opposite sign and thus is ambiguously signed.

Thus the GM conditions, necessary for signing all diagonal elements in A under the stability hypothesis, themselves lead to problems with signing off-diagonal cofactors, when both positive and negative cycles appear in A.

6.5 SCOPE OF THE CORRESPONDENCE PRINCIPLE

With this preliminary work out of the way, we can return to the general question raised in this chapter, namely, What is the scope of the correspondence principle in the qualitatively specified economic environment? Alternatively, for what class of qualitative matrices A and qualitative vectors b does the hypothesis of stability of A allow the system $Ax = b$ to be solved for the sign pattern of x (full sign solvability) or the sign pattern of a subset of the entries of x (partial sign solvability)?

As noted earlier, the fact that the potential stability problem remains unsolved means that we have only partial answers to this question. Theorem 6.6 below gives necessary and sufficient conditions for full sign solvability under the stability hypothesis (when A has a negative diagonal) in the case of a "full" matrix (no zeros in A or b). Theorem 6.7 provides sufficient conditions for full sign solvability for stable negative diagonal matrices A in the special case where the b vector contains only one nonzero element. Theorem 6.8 gives necessary and sufficient conditions for full and partial sign solvability in the GM case.

THEOREM 6.6. *Let A be an $n \times n$ full matrix ($a_{ij} \neq 0$ for all $i, j = 1, \ldots, n$) with all diagonal entries negative. Then, for $n > 2$, A^{-1} is signed under the stability hypothesis if and only if A is a Metzler or Morishima matrix. Further, if both A and b are full, $Ax = b$ is fully sign solvable under the stability hypothesis if and only if A is Metzler or Morishima, and b satisfies the sign conditions of Theorem 6.3. This holds for all n.*

PROOF. Note that A^{-1} signed under the stability hypothesis implies that A is a GM-matrix, by Theorem 6.4. By Lemma 6.1, if A is a GM matrix containing both a positive and a negative cycle with an index in common, then A^{-1} is not signed under the stability hypothesis. Full matrices containing both positive and negative cycles satisfy this condition. Thus if A^{-1} is to be signed under the stability hypothesis, either A has no positive cycles or A has no negative cycles. But for $n > 2$, if A is full, then A must contain at least one positive cycle. Suppose that $a_{12}a_{23}a_{31} < 0$, and $a_{21}a_{13}a_{32} < 0$. Thus at least one cycle of length two, $a_{12}a_{21}$, $a_{23}a_{32}$, or $a_{13}a_{31}$, is positive. Hence A must be either a Metzler or a Morishima matrix, with A^{-1} of appropriate sign pattern. With b

full, $Ax = b$ is sign solvable only if A^{-1} is signed under the stability hypothesis and the sign conditions of Theorem 6.3 guarantee sign solvability in this case. For $n = 2$, if A is a sign stable matrix, then A^{-1} is signed, even though A has a negative cycle. On the other hand, if A and b are full, then $Ax = b$ is not fully sign solvable in this case, as is easily verified.

THEOREM 6.7. *Let A be an $n \times n$ matrix with negative diagonal entries, and let $b_k \neq 0$ be the unique nonzero entry in b. Then $Ax = b$ is fully sign solvable under the stability hypothesis if*

(i) *All positive cycles in A contain the index k, and for any $i \neq k$, all nonzero paths $a(i \to k)$ have the same sign; or*
(ii) *A is a GM-matrix, satisfying the path condition in (i).*

PROOF. The system $Ax = b$ is fully sign solvable in this case if all cofactors $A_{k1}, A_{k2}, \ldots, A_{kn}$ are signed under the stability hypothesis. If condition (i) is satisfied, then all principal minors appearing in any of these cofactors satisfy the Hicks conditions, since all nonzero cycles appearing in these cofactors are negative. This signs A_{kk} as $(-1)^{n-1}$, and the path condition then signs the other cofactors. If condition (ii) is satisfied, stability of a GM-matrix implies the Hicks conditions, so that all principal minors of A are signed, including A_{kk}, which has sign $(-1)^{n-1}$. Again, the path condition signs all the off-diagonal cofactors in the kth column of A^{-1}, and $Ax = b$ is sign solvable.

THEOREM 6.8. *Given that A is an $n \times n$ GM matrix, the comparative statics system $Ax = b$ is fully sign solvable under the stability hypothesis if and only if $b_k \neq 0$ implies that all nonzero paths $a(i \to k)$ have the same sign for any $i, i = 1, \ldots, n$, and that, for any i, $A_{ki} b_k$ has weakly the same sign for all $k = 1, \ldots, n$. $Ax = b$ is partially sign solvable under the stability hypothesis for x^1 a subset of x if and only if $b_k \neq 0$ implies that for any i an index of a member of the set x^1, all nonzero paths $a(i \to k)$ have the same sign, and that, for any such i, $A_{ki} b_k$ has weakly the same sign for all $k = 1, \ldots, n$.*

PROOF. Immediate from Corollary 1 to Theorem 6.1 and Theorem 6.3.

It is easy to verify that if all nonzero paths $a(i \to j)$ in a GM-matrix are positive, then, for $B \in Q_A^*$, $B_{ji}/\det B < 0$, and conversely when all nonzero paths $a(i \to j)$ are negative. Since diagonal elements in B^{-1} are signed negative under the stability hypothesis, this gives rise to the following rules for full and partial sign solvability under the stability hypothesis, in the GM case:

(1) If $b_k \neq 0$ for a unique element in b, then the necessary and sufficient condition for full sign solvability under the stability

hypothesis is that, for any $i \neq k$, all nonzero paths $a(i \to j)$ have the same sign, and there is partial sign solvability for x_k independent of the paths condition, and for $x_i \neq x_k$ if the path condition is satisfied for that i.

(2) If $b_1, b_2, \ldots, b_t$ are the only nonzero entries in b, with $\operatorname{sgn} b_1 = \operatorname{sgn} b_2 = \cdots = \operatorname{sgn} b_r$, and sign $b_{r+1} = \operatorname{sgn} b_{r+2} = \cdots = \operatorname{sgn} b_t \neq \operatorname{sgn} b_1$, then the necessary and sufficient conditions for full sign solvability are that (i) either $a(i \to j) \geq 0$ for $i \neq j$, $i, j = 1, \ldots, r$, and $a(i \to j) \leq 0$ for $i \neq j, i, j = r+1, \ldots, t$, or the opposite pattern holds; and (ii) for $i = t+1, \ldots, n$, all nonzero paths $a(i \to j)$ are of the same sign, for $j = 1, \ldots, r$, opposite to that of all nonzero paths $a(i \to j)$ for $j = r+1, \ldots, t$.

There is partial sign solvability for x_i if (i) holds for $i \in \{1, \ldots, t\}$ or if (ii) holds for $i \in \{t+1, \ldots, n\}$.

Examples to illustrate these results are given below.

EXAMPLE 1. A has no positive cycles (L-matrix or SNS matrix).

$$\begin{bmatrix} - & + & 0 & 0 \\ - & - & + & 0 \\ 0 & - & - & + \\ 0 & 0 & - & - \end{bmatrix} \begin{bmatrix} x_1 \\ x_2 \\ x_3 \\ x_4 \end{bmatrix} = \begin{bmatrix} + \\ 0 \\ 0 \\ 0 \end{bmatrix}$$

$$\Rightarrow \begin{bmatrix} x_1 \\ x_2 \\ x_3 \\ x_4 \end{bmatrix} = \begin{bmatrix} - & - & + & - \\ + & - & - & - \\ - & + & - & - \\ + & - & + & - \end{bmatrix} \begin{bmatrix} + \\ 0 \\ 0 \\ 0 \end{bmatrix} = \begin{bmatrix} - \\ + \\ - \\ + \end{bmatrix}$$

EXAMPLE 2. A has no negative cycles (Metzler matrices).

$$\begin{bmatrix} - & + & 0 & 0 \\ + & - & + & 0 \\ 0 & + & - & + \\ 0 & 0 & + & - \end{bmatrix} \begin{bmatrix} x_1 \\ x_2 \\ x_3 \\ x_4 \end{bmatrix} = \begin{bmatrix} + \\ + \\ + \\ + \end{bmatrix}$$

$$\Rightarrow \begin{bmatrix} x_1 \\ x_2 \\ x_3 \\ x_4 \end{bmatrix} = \begin{bmatrix} - & - & - & - \\ - & - & - & - \\ - & - & - & - \\ - & - & - & - \end{bmatrix} \begin{bmatrix} + \\ + \\ + \\ + \end{bmatrix} = \begin{bmatrix} - \\ - \\ - \\ - \end{bmatrix}$$

EXAMPLE 3. GM-matrix with positive and negative cycles.

$$\begin{bmatrix} - & + & 0 & - \\ - & - & + & 0 \\ 0 & - & - & + \\ + & 0 & - & - \end{bmatrix}\begin{bmatrix} x_1 \\ x_2 \\ x_3 \\ x_4 \end{bmatrix} = \begin{bmatrix} + \\ 0 \\ 0 \\ 0 \end{bmatrix}$$

$$\Rightarrow \begin{bmatrix} x_1 \\ x_2 \\ x_3 \\ x_4 \end{bmatrix} = \begin{bmatrix} - & ? & - & ? \\ ? & - & ? & - \\ - & ? & - & ? \\ ? & - & ? & - \end{bmatrix}\begin{bmatrix} + \\ 0 \\ 0 \\ 0 \end{bmatrix} = \begin{bmatrix} - \\ ? \\ - \\ ? \end{bmatrix}$$

Note that the only positive cycles in A have index sets that contain all indices in A, which appear in one or more negative cycles in A such that all rows and columns of A have ambiguously signed elements.

EXAMPLE 4. GM-matrix with full sign solvability.

$$\begin{bmatrix} - & + & 0 \\ - & - & + \\ + & 0 & - \end{bmatrix}\begin{bmatrix} x_1 \\ x_2 \\ x_3 \end{bmatrix} = \begin{bmatrix} 0 \\ + \\ + \end{bmatrix} \Rightarrow \begin{bmatrix} x_1 \\ x_2 \\ x_3 \end{bmatrix} = \begin{bmatrix} - & - & - \\ ? & - & - \\ - & - & - \end{bmatrix}\begin{bmatrix} 0 \\ + \\ + \end{bmatrix} = \begin{bmatrix} - \\ - \\ - \end{bmatrix}$$

6.6 INTERPRETATION

The question arises as to the economic interpretation of the conditions characterizing the GM class of matrices in the setting of simultaneous market clearing of a collection of markets. Let (P) (for positive cycles) denote the Hicksian rule, Substitutes of substitutes and complements of complements are substitutes, while substitutes of complements and complements of substitutes are complements, combined with (weak) symmetry of substitutability-complementary relations; i.e., i a substitute for j implies that j is a (weak) substitute for i, and similarly for complements. And let (N) (for negative cycles) denote the converse rule, Substitutes of substitutes and complements of complements are complements, while substitutes of complements and complements of substitutes are substitutes, combined with (weak) antisymmetry of substitutability-complementary relations; i.e., i a substitute for j implies that j is a (weak) complement for i, and similarly for complements.

Under either (P) or (N), assume that the law of demand holds for all goods. Then both matrices for which (P) holds and those for which (N) holds are members of the GM class, so that dynamic stability implies Hicksian stability. The comparative statics results for (N) are given in Theorem 6.1, and those for (P) are given in Theorem 6.3. Moreover, both (P) and (N) are consistent with the competitive assumptions, assuming the properties of the *numeraire* commodity are appropriately chosen, per the Sonnenschein (1973) result.

Further, the GM conditions imply that in the case where an economy contains both (P) and (N) relationships, if i has both a direct or indirect (N) relationship with j and a (P) relationship with j, then the (P) relationship with j has to involve all the commodities that are in i's (N) relationship with j. If this condition is violated, then, by Theorem 6.4, there exist matrices of this qualitative class for which dynamic stability does not imply Hicksian stability. The relevance of Hicksian stability in addressing comparative statics questions under the correspondence principle is clear from the discussion in this chapter.

7

The Competitive Equilibrium: Comparative Statics

7.1 INTRODUCTION

Equilibrium is a recurring theme in scientific explanation. Observed physical, chemical, ecological, and economic processes are often interpreted as a balance of interacting influences, formally described by equation (1.1) in Chapter 1. Some equilibria are unchanged for finite time periods; others evolve as described by the equilibrium equations. Depending upon context, the equilibrium equations can themselves impose restrictions on how observations can change in response to changes in the environment and over time.

This chapter and the next show how the equilibrium hypothesis can be exploited to yield definite qualitative conclusions. Our example is the competitive equilibrium. The competitive equilibrium imposes certain weak restrictions upon observed data. First, observed quantities (excess demands for commodities) are not affected by rescaling prices (the homogeneity postulate). Second, a kind of conservation of force ensures that a proper measure of the pressures pushing and pulling on equilibrium in individual markets has no net weight (Walras' law). These restrictions, singly and together, imply definite restrictions on observed data. Other subject areas might also provide equilibrium restrictions that yield testable conclusions in a qualitative environment.

In this chapter, we first present the basic competitive model, and then derive those comparative statics theorems that can be established for the competitive model given a qualitative framework. In Chapter 8, we examine the stability of the competitive equilibrium in a qualitative setting. The major difference in approach between these chapters and the earlier chapters is that the assumptions underlying the competitive model provide restrictions on the signs of row and column sums of qualitative matrices.

7.2 A COMPETITIVE ECONOMY

We consider a private ownership competitive economy in which there are m consumers, $n+1$ commodities, and l firms. The economy is competitive in the sense that given a price vector $p=(p_0,\dots,p_n)$, both consumers and firms treat prices as parameters of their decision problems; i.e., consumers and firms ignore the effects of their actions on prices. By a private ownership economy we mean an economy in which title to all resources is held by consumers, who also share in profits of firms according to their stock ownership in firms.

It is assumed the preferences of the ith consumer can be represented by a real-valued utility function u^i defined over commodity bundles x^i, where $x^i=(x_{i0},x_{i1},\dots,x_{in})$, with x_{ij} $j=0,\dots,n$ denoting the number of units of commodity j allocated to consumer i. If $x_{ij}>0$, then commodity j is consumed by consumer i, while if $x_{ij}<0$, commodity j is supplied as a labor service in the production process by consumer i. Commodity bundles for consumer i are constrained to lie within a consumption set X^i, a subset of Euclidean $(n+1)$-dimensional space, where X^i is determined by such physical considerations as the impossibility of supplying more than twenty four hours of labor per day, etc. Holdings of resource j by consumer i, are denoted by $\mathring{x}^i_j$ with $\mathring{x}_i=(\mathring{x}_{i0},\mathring{x}_{i1},\dots,\mathring{x}_{in})$ denoting the vector of resource holdings by consumer i. Thus if $\mathring{x}_j$ is the total existing stock of resource j, the private ownership assumption implies

$$\mathring{x}_j=\sum_{i=1}^{m}\mathring{x}_{ij}.$$

Similarly, the kth firm may be characterized by an input-output vector $y^k=(y_{k0},y_{k1},\dots,y_{kn})$, where y_{kj} $j=0,\dots,n$ denotes the number of units of commodity j produced by firm k if $y_{kj}>0$, and denotes the number of units of commodity j used as an input in production by firm k if $y_{kj}<0$. The technology of the economy determines a production set Y^k for the kth firm that specifies the set of physically feasible input-output vectors. For example $y^k>0$ is excluded from Y^k since this would represent positive outputs with zero inputs.

In terms of the above notation, one may define a feasible state of the economy as a vector $(x^1,x^2,\dots,x^m,y^1,y^2,\dots,y^l)$ such that

$$x^i\in X^i\ i=1,\dots,m;\ y^k\in Y^i\ k=1,\dots,l;\ \text{and} \tag{1}$$

$$\sum_{i=1}^{m}x_{ij}-\sum_{k=1}^{1}y_{kj}-\sum_{i=1}^{m}\mathring{x}_{ij}\le 0\ j=0,\dots,n. \tag{2}$$

Condition (1) guarantees physical feasibility of commodity bundles and input-output vectors, while condition (2) specifies overall balance in the economy in the sense that consumption of any commodity cannot exceed production of the commodity plus the amounts held in resource stocks.

It is further postulated that the kth firm acts to maximize profits, i.e., that firm k chooses an input-output vector y^k (if such a vector exists) to maximize profits π^k, where $\pi^k = \sum_{j=0}^{n} p_j y_{kj}$ subject to $y^k \in Y^k$, where under the competitive assumption, prices are treated as constants by the kth firm. The resulting profits are distributed to consumers according to their stock ownership. Let α_{ik} denote the percentage of firm k's stock held by consumer i. Then $\alpha_{ik} \geq 0$, $\sum_{i=1}^{m} \alpha_{ik} = 1$ for $k = 1, \ldots, l$, and the amount of firm k's profits assigned to consumer i is $\alpha_{ik}\pi^k$.

The ith consumer is assumed to maximize utility subject to his budget constraint; i.e., consumer i chooses x^i (if such a vector exists) to maximize $u^i(x^i)$ subject to

$$x^i \in X^i$$

and

$$\sum_{j=0}^{n} p_j x_{ij} \leq \sum_{j=0}^{n} p_j \mathring{x}_{ij} + \sum_{k=1}^{1} \alpha_{ij}\pi^k,$$

where p is taken as a constant vector. Thus the ith consumer is constrained by the fact that the net value of her purchases cannot exceed the value of her resource holdings plus the profits she obtains from ownership of stock in firms.

Given the set of resource endowments $\mathring{x}^1, \ldots, \mathring{x}^m$ and the stock ownership matrix $[\alpha_{ij}]$, we denote, for any p, the profit-maximizing input-output vector of firm k by $y^k(p)$, and similarly, we denote the utility-maximizing commodity bundle of consumer i by $x^i(p)$. Let $E_j(p)$ denote excess demand for commodity j, where

$$E_j(p) = \sum_{i=1}^{m} x_{ij}(p) - \sum_{k=1}^{1} y_{kj}(p) - \sum_{i=1}^{m} \mathring{x}_{ij}.$$

Hence excess demand for commodity j at the price vector p is the difference between demands for commodity j by utility-maximizing consumers and the amount of j available, which equals that being supplied by profit-maximizing firms plus the existing stock.

At a utility-maximizing position, budget constraints of consumers must be satisfied. If we further assume that no consumer has a "bliss point" (i.e., a preference such that, for any i, given any $x^i \in X^i$, there is an $x^{*i} \in X^i$ that is preferred to x^i), then budget constraints for all consumers are satisfied as equalities. Summing the budget constraints over all consumers and evaluating at that point where consumers maximize utility and firms maximize profits, we obtain

$$\sum_{i=1}^{m} \left(\sum_{j=0}^{n} p_j x_{ij}(p) - \sum_{j=0}^{n} p_j \mathring{x}_{ij} - \sum_{k=1}^{1} \alpha_{ij} \sum_{j=0}^{n} p_j y_{kj}(p) \right) = 0.$$

Using the fact that $\Sigma_{i=1}^{m} \alpha_{ij} = 1$ and reversing the order of summation, we obtain Walras's law:

$$\sum_{j=0}^{n} p_j E_j(p) = 0$$

for any price vector p. In addition the competitive assumption of parametric prices implies that $E_j(p)$ is (positively) homogeneous of degree zero in p. This follows because $u^i(x^i)$ is independent of p. Multiplying all prices by a positive constant does not change the budget constraint for any consumer. Hence the utility-maximizing bundle does not change. Similarly, multiplying prices by a positive constant does not change the profit-maximizing input-output vector for any firm (although the amount of profits earned will generally change). Hence a second property of excess demand functions is (positive) homogeneity of degree zero:

$$E_j(\lambda p) = E_j(p) \text{ for } \lambda > 0 \; j = 0, \ldots, n.$$

Given the private ownership parameters of the economy, i.e., the resource endowment vectors $\mathring{x}^1, \ldots, \mathring{x}^m$ and the stock ownership matrix $[\alpha_{ik}] i = 1, \ldots, m,\ k = 1, \ldots, l$, a competitive equilibrium is defined as an $(m + l + 1)$-tuple $\{p^*; x^1(p^*), \ldots, x^m(p^*); y^1(p^*), \ldots, y^l(p^*)\}$ such that

(i) $x^i(p^*)$ maximizes $u^i(x^i)$ subject to $x^i \in X^i$, and

$$\sum_{j=0}^{n} p_j^* x_{ij} \leq \sum_{j=0}^{n} p * \mathring{x}_{ij} + \sum_{k=1}^{1} \alpha_{ik} \pi^k \qquad i = 1, \ldots, m;$$

(ii) $y^k(p^*)$ maximizes $\pi^k = \Sigma_{j=0}^{n} p_j^* y_{kj}$ subject to $y^k \in Y^k$ $k = 1, \ldots, l$; and

(iii) $E_j(p^*) \leq 0 \; j = 0, \ldots, n$.

Thus a competitive equilibrium is a price vector p^* and a state of the economy such that the utility-maximizing choices of consumers and profit-maximizing choices of firms generate a feasible state of the economy, where excess demand for each commodity is nonpositive. The analysis in this chapter assumes a competitive equilibrium. It does not deal with the complex problem of specifying an axiom system sufficient to guarantee the existence of a competitive equilibrium, including the pioneering work of Wald (1951) and later contributions by Arrow and Debreu (1954), Debreu (1959), Gale (1969), and McKenzie (1960a).

Our concern in this chapter is with the comparative statics properties of the competitive equilibrium, within a qualitative framework. In what follows, we will make the simplifying assumptions that at competitive equilibrium positions, excess demand for each commodity is zero (rather than merely nonpositive) and equilibrium price vectors are strictly positive. Thus we exclude the case of "free goods" at equilibrium. It also will be assumed that excess demand functions are differentiable, which implies among other things, strict convexity of production sets and of preferences.

It is convenient to assume that good #0 is an "always desired good," i.e., that $E_0(p) = +\infty$ if $p_0 = 0$. Under this assumption, good #0 may be chosen as *numeraire*; i.e., the price of good #0 may be fixed at 1. Under homogeneity, $E_j(\lambda p) = E_j(p)$ for every $\lambda > 0$; hence, taking the domain of excess demand functions as the set $(1, p_1, \ldots, p_n)$, with $p_i \geq 0$ $i = 1, \ldots, n$, rather than the set $(p_0, p_1, \ldots, p_n)$, with $p_0 > 0$ $p_i \geq 0$ $i = 1, \ldots, n$, constitutes no restriction. Under this normalization, the price of good i may now be interpreted as the number of units of the *numeraire* commodity that exchanges for one unit of good i. Another way of summarizing this property of the competitive model is the statement that at equilibrium, only relative prices, not absolute prices, are determined.

We also note that because of Walras's law, $\Sigma_{j=0}^{n} p_j E_j(p) = 0$ for every p, so excess demands are not independent; in particular, given the choice of good #0 as *numeraire*, we have $\Sigma_{j=1}^{n} p_j E_j(p) = E_0(p)$ for every p. Hence excess demand for the *numeraire* commodity is determined once excess demands for the remaining n commodities are determined.

7.3 SOME HISTORICAL COMMENTS

From the point of view of the history of economic theory, the main purpose underlying the construction of the competitive model was not to derive comparative statics theorems, but instead to arrive at welfare judgments concerning a market economy. The classical results concerning this are the two basic welfare theorems of economics: (1) every competitive equilibrium is a Pareto optimum; and (2) every Pareto optimum can be achieved as a competitive equilibrium, under an appropriate reallocation of resources among consumers. (A Pareto optimum is an allocation of goods and services among consumers such that it is not possible to make one consumer better off except by making some other consumer worse off.) These welfare theorems hold under very general conditions (the main qualifications are the assumptions that externalities and monopoly power are absent in the economy). Because of the desire for generality of welfare results, the competitive model has been formulated in very general terms as well. In fact, as Sonnenschein (1973) has shown, any set of continuous functions satisfying Walras's law and homogeneity can be derived as the excess demand functions for a competitive economy by making an appropriate choice of utility functions for consumers.

In comparative statics, the idea is to come up with predictive statements about the economy, and the more restrictive the statement, the better, because it makes it easier to test the validity of the comparative statics prediction. The consequence is that in developing a comparative statics of competitive general equilibrium models, restrictive assumptions on the model are a necessity. Without such restrictive assumptions, the competitive model predicts only that Walras's law, homogeneity, and continuity will characterize excess demands.

In the neoclassical economics literature, there are only two cases in which restrictive comparative statics theorems have been derived concerning qualitatively specified general equilibrium models. Both derive originally from Hicks's *Value and Capital* (1939). Important work, as noted earlier, was done later by Metzler, Mosak, and Morishima, among others. First, Hicks examined what has come to be known as the case of a gross substitute economy, i.e., an economy in which all goods are gross substitutes for one another. In such an economy, an increase in the price of any good i leads to a decrease in the excess demand for good i and an increase in the excess demand for all other goods.

Hicks formulated his famous three laws of comparative statics for a gross substitute economy:

1. A shift in excess demand from *numeraire* to good i increases the equilibrium price of good i.
2. This also increases the equilibrium prices of all other non-*numeraire* goods in the economy.
3. Such a shift in excess demand increases the equilibrium price of good i proportionately more than it increases the prices of the other goods in the economy.

Hicks derived these laws under the assumption that the economy was stable in his sense of term, i.e., Hicksian stable. He also derived them for only small-dimension cases. As it turns out, the Hicksian laws apply to an economy with an arbitrary number of goods, as long as the gross substitute assumptions hold; and Walras's law and homogeneity guarantee that a gross substitute economy will satisfy Hicksian stability as well.

The other case that Hicks examined in *Value and Capital* was the somewhat peculiar case of an economy in which relationships among goods obeyed the rule that substitutes of substitutes and complements of complements are substitutes, while substitutes of complements and complements of substitutes are complements. Again for small-dimension cases, Hicks showed that if an economy obeying this rule also satisfies Hicksian stability, then the following rules also hold:

1. A shift in excess demand from *numeraire* to good i increases the equilibrium price of good i.
2. Such a shift increases the equilibrium price of all substitutes and decreases the equilibrium price of all complements.

Morishima's later work extended Hicks's argument to the case of an n-commodity world. However, in contrast to the case of a gross substitute world, there is no guarantee that a Morishima economy will satisfy Hicksian stability. In fact, it turns out that if all goods in an economy obey the Hicks-Morishima conditions, then the economy is not Hicksian stable and not dynamically stable either. It requires quantitative conditions to guarantee Hicksian stability in this kind of a world.

One interesting feature of these two cases is that they are qualitative cases. In the rest of this chapter, we will be involved in seeing the extent to which the use of qualitative models, specified in terms of complementarity-substitutability relations among goods, can extend the scope of comparative statics results beyond those early results achieved by Hicks, Metzler, and Morishima.

7.4 RESTATEMENT OF SIGN SOLVABILITY

The comparative statics properties of the competitive equilibrium relate to changes that occur in the equilibrium price vector following a change in the environmental parameters of the model (e.g., preferences, resource endowments, distribution of stock ownership, technology, etc.). In comparative statics analysis, the existence of an equilibrium for various choices of the environmental parameters is posited. Let β denote an environmental parameter, and let β', β'' be two values of this parameter, with $p^*(\beta')$ and $p^*(\beta'')$ being the corresponding equilibrium price vectors so that $E_j[p^*(\beta'), \beta'] = 0$, $E_j[p^*(\beta''), \beta''] = 0$ for $j = 1, \dots, n$. As in the earlier chapters, we will be concerned only with infinitesimal changes in β, under the assumption that E_j is differentiable. Hence the comparative statics problem we seek to solve is of the form

$$\sum_{i=1}^{n} \frac{\partial E_j}{\partial p_i} \frac{dp_i^*}{d\beta} = -\frac{\partial E_j}{d\beta} \qquad j = 1, \ldots, n,$$

where $\partial E_i/\partial p_i$ and $\partial E_j/\partial \beta$, $i, j = 1, \ldots, n$, are evaluated at the equilibrium position, and $dp_i^*/d\beta$ $i = 1, \ldots, n$ are the changes in equilibrium prices of the non-*numeraire* goods due to a change in the environmental parameter β, and are the unknowns of the comparative statics problem. Note that the *numeraire* good, good #0, is excluded from consideration since its price is fixed at 1, independently of the value that β takes on.

Since we are concerned primarily with comparative statics "in the small," the implications of homogeneity and of Walras's law in differential form are of interest. By Euler's theorem, if $f(x_1, \ldots, x_n)$ is homogeneous of degree k in the x's, then

$$\sum_{i=1}^{n} \frac{\partial f}{\partial x_i} x_i = kf.$$

Hence the property of homogeneity of degree zero in prices of excess demand functions may be stated as follows:

$$\sum_{r=0}^{n} \frac{\partial E_j}{\partial p_r} p_r = 0 \; j = 0, \ldots, n$$

is an identity in p.

Similarly, differentiate Walras's law with respect to any price p_r to obtain

$$\sum_{j=0}^{n} \frac{\partial E_j}{\partial p_r} p_j + E_r(p) = 0 \qquad r = 0, \ldots, n.$$

Evaluating at equilibrium prices p^*, where $E_r(p^*) = 0$ $r = 0, \ldots, n$, we have

$$\sum_{j=0}^{n} \frac{\partial E_j}{\partial p_r} p_j^* = 0 \qquad r = 0, \ldots, n.$$

Suppose that all equilibrium prices are positive and that we now choose the units of measurement of all non-*numeraire* commodities so that $p_j^* = 1$, $j = 1,\ldots,n$. For simplicity of notation, let a_{jr} denote $\partial E_j/\partial p_r$ (evaluated at p^*). Then the differential condition for Walras's law (W) can be written as

$$\sum_{r=0}^{n} a_{jr} = 0 \qquad j = 0,\ldots,n,$$

and the differential condition for homogeneity (H) can be written as

$$\sum_{j=0}^{n} a_{jr} = 0 \qquad r = 0,\ldots,n.$$

Further, let $A^* = [a_{jr}]j,\ r = 0,\ldots,n$; $A = [a_{jr}]j,\ r = 1,\ldots,n$. Then (W) and (H) imply that row and column sums in A^* are all zero. The local comparative statics problem of the competitive equilibrium (C) can be formulated as follows: Solve for $dp/d\beta$ in

$$\text{(C)} \quad A\frac{dp}{d\beta} = b,$$

where $dp/d\beta = [dp_j/d\beta]j = 1,\ldots,n$, $b = -[\partial E_j/\partial\beta]j = 1,\ldots,n$, and $A^* = [a_{ij}]i, j = 0,\ldots,n$ satisfies the condition that row and column sums be zero. Note that it is $A = [a_{ij}]$, $i, j = 1,\ldots,n$ rather than A^* that appears in the comparative statics problem.

Qualitative analysis of (C) is concerned with determining the class of qualitatively specified competitive systems, as described by A^* and b, under which it is possible to determine the sign pattern of $dp/d\beta$ in (C), given the quantitative restrictions imposed on A^* by (W) and (H). By a qualitatively specified competitive system A^*, we mean one in which the substitutability-complementarity relations among goods are specified, but not the quantitative strength of such relationships. As in the previous chapters, we use the notation $Q_{A^*} = \{C^* \mid \operatorname{sgn} C^* = \operatorname{sgn} A^*\}$ to describe a qualitative matrix consisting of all $(n+1)\times(n+1)$ matrices with the same sign pattern as A^*. Any asterisked matrix, such as C^*, is of dimension $(n+1)\times(n+1)$, and the corresponding matrix without an asterisk involves the deletion of the 0th row and column; i.e., $C^* = [c_{ij}]i,\ j = 0,\ldots,n$, and $C = [c_{ij}]i,\ j = 1,\ldots,n$. We also define the set S_{A^*} as

$$S_{A^*} = \left\{C^* \mid C^* \in Q_{A^*} \text{ and } \sum_{j=0}^{n} c_{ij} = 0\ i = 1,\ldots,n,\ \sum_{i=0}^{n} c_{ij} = 0\ j = 0,\ldots,n\right\},$$

while

$$V_{A^*} = \left\{ C^* \mid C^* \in Q_{A^*} \text{ and } \sum_{i=0}^{n} c_{ij} = 0 \; j = 0, \ldots, n \right\}.$$

S_{A^*} is the set of $(n+1) \times (n+1)$ matrices that have the same sign pattern as A^* and also satisfy the condition that row and column sums are all zero, i.e., satisfy (W) and (H). V_{A^*} is a larger set of $(n+1) \times (n+1)$ matrices that have the same sign pattern as A^* but satisfy only the condition that the column sums of such matrices be zero, i.e., (W) but not necessarily (H).

The reason for introducing V_{A^*} is that it is easier to analyze the comparative statics properties of a system in which only column sum (or only row sum) restrictions apply, as compared to the case where both row and column sum restrictions are present, as in S_{A^*}. There is also some limited independent interest in the V_{A^*} case on purely economic grounds, e.g., when consumers suffer from "money illusion," as when prices enter consumer utility functions. In any case, the approach adopted in this chapter is to first examine the comparative statics properties of a competitive economy operating with only (W) as a restriction. Following this, the more interesting case of an economy operating under both (W) and (H) is taken up.

Using this notation, we introduce the following concepts:

1. A is said to have *signed determinant under* (W) if B^*, $C^* \in V_{A^*}$ implies $\text{sgn}|B| = \text{sgn}|C|$.
2. A is said to have *signed inverse under* (W) if B^*, $C^* \in V_{A^*}$ implies $\text{sgn}\, B^{-1} = \text{sgn}\, C^{-1}$.
3. $Ax = b$ is said to be *fully sign solvable under* (W) if B^*, $C^* \in V_{A^*}$, $d, e \in Q_b$; $By = d$; and $C_z = e$ implies $y \in Q_z$.
4. $Ax = b$ is said to be *partially sign solvable under* (W) if some subset of x is sign solvable under (W).
5. A is said to have *signed determinant under* (W) *and* (H) if B^*, $C^* \in S_{A^*}$ implies $\text{sgn}|B| = \text{sgn}|C|$.
6. A is said to have *signed inverse under* (W) *and* (H) if B^*, $C^* \in S_{A^*}$ implies $\text{sgn}\, B^{-1} = \text{sgn}\, C^{-1}$.
7. $Ax = b$ is *fully sign solvable under* (W) *and* (H) if B^*, $C^* \in S_{A^*}$; $d, e \in Q_b$; $By = d$; and $C_z = e$ implies $y \in Q_z$.
8. $Ax = b$ is *partially sign solvable under* (W) *and* (H) if a subset of x is sign solvable under (W) and (H).

7.5 SIGN SOLVABILITY UNDER WALRAS'S LAW

It follows immediately from the definitions above that, since $S_{A^*} \subseteq V_{A^*}$, if A has signed determinant under (W) (signed inverse under (W), etc.), then A has signed determinant under (W) and (H) (signed inverse under (W) and (H), etc.). Thus, in what follows, we will be deriving sufficient conditions for qualitative solvability under (W) and (H). In this analysis, the concept of a free entry in A under (W) is of interest.

DEFINITION 7.1. *a_{ij} is a free entry in A under (W) if, given any real number $M > 0$, there exists $C^* \in V_{A^*}$ such that $|c_{ij}| > M|c_{rs}|$ for every $r \neq i$ or $s \neq jr, s = 1, \ldots, n$.*

Intuitively, a free entry in A under (W) is an entry that can be made arbitrarily large in absolute value relative to all other entries in A without violating either (W) or the sign pattern of A^*. For example, assume A^* has sign pattern

$$\begin{bmatrix} - & + & + \\ + & - & + \\ + & + & - \end{bmatrix},$$

with rows and columns numbered 0, 1, 2. Then a_{11} is a free entry in A under (W). To see this, consider any $B^* \in V_{A^*}$. Let $\hat{b}$ denote the absolute value of the largest entry in B, excluding b_{11}. Then choose C^* such that all entries in C^* are identical to those in B^* except that $c_{11} = -M\hat{b}$, $(M > 1)$ and choose $c_{01} = M\hat{b} - \Sigma_{i=2}^{n} b_{i1}$. Then $c_{01} > 0$, and all column sums in C^* are zero. Moreover, $|c_{11}| \geq M|c_{ij}|$ for every entry c_{ij} in C. In contrast, a_{21} is not a free entry in A under (W) since by the column sum restriction together with $a_{01} > 0$, we have $|c_{21}| = c_{21} < |c_{22}|$ for every $C^* \in V_{A^*}$.

It follows from the definition of a free entry that a_{ij} is a free entry in A under (W) if and only if $a_{ij}a_{0j} < 0$. In addition, the following properties of free entries are easily established:

(i) if A_{ij} is the $(n-1) \times (n-1)$ cofactor of a_{ij} in A, then a_{ij} a free entry in A under (W), $A_{ij} \neq 0$ implies that if A has signed determinant under (W), $\text{sgn}|A| = \text{sgn}\, a_{ij}A_{ij}$; and

(ii) if A has signed determinant under (W), then $|A|$ has the same sign as the sign of any nonzero term in the expansion of $|A|$ involving $n-1$ free entries.

Strong Dependence: Full A^* Matrices

In applying qualitative analysis to the case where column sums in A^* are zero, we first consider the "strong dependence" case, where A^* is "full," i.e., the case in which every entry in A^* is nonzero (see Quirk [1968a]). Because $a_{ij}a_{0j} < 0$ when a_{ij} is a free entry, and because $\sum_{i=0}^{n} a_{ij} = 0$, $j = 0,\dots,n$, the assumption that every entry in A^* is nonzero implies that every column in A^* contains at least one free entry. The further implications of the assumption that A has signed determinant under (W) are derived in the following lemmas.

LEMMA 7.1. *Assume A^* satisfies $a_{ij} \neq 0$, $i, j = 0,\dots,n$. If A has signed determinant under (W), then there exists a term in the expansion of $|A|$ containing n free entries in A under (W).*

PROOF. Every column of A contains at least one free entry; hence, reindex a free entry in the first column of A into a_{11}. If there exists a free entry in the remaining $n-1$ rows and columns of A, reindex into a_{22}. If not, then $a_{12},\dots,a_{1n}$ represent the unique free entries in columns $2,\dots,n$; hence, in $[A_{11}]$, the submatrix of A with row 1 and column 1 deleted, the jth column consists entirely of entries all of the same sign opposite to that of a_{1j} for $j = 2,\dots,n$. But then each of these entries is free in $[A_{11}]$; i.e., if we choose a_{1j} and a_{0j} sufficiently large in absolute value, there exists $C^* \in V_{A^*}$ such that $|c_{ij}| > M|c_{rs}|$ for any given i, $j = 2,\dots,n$, $r \neq i$ and $s \neq j$ $r, s = 2,\dots,n$. Thus $[A_{11}]$ contains $n-2$ columns of free entries, all of the same sign in any one column. For any assigned signs, A_{11} will have at least two terms consisting solely of free entries but of opposite sign, so A_{11} is not signed under (W); thus by property (i) of free entries, A does not have signed determinant under (W). By a similar argument, it is possible to reindex free entries into $a_{22}, a_{33},\dots,a_{n-1,n-1}$, so that by property (ii) of free entries, $\operatorname{sgn}|A| = \operatorname{sgn} a_{11}a_{22}\cdots a_{n-1,n-1}a_{nn}$. If a_{nn} is not a free entry, there exists an entry a_{in} that is free, where $a_{in}a_{nn} < 0$. Reindex this entry into a_{1n}. Then $\operatorname{sgn}|A| = \operatorname{sgn} a_{n1}a_{22}\cdots a_{n-1,n-1}a_{1n}$, which implies $\operatorname{sgn} a_{11}a_{nn} = -\operatorname{sgn} a_{n1}a_{1n}$, or $\operatorname{sgn} a_{11} = \operatorname{sgn} a_{n1}$, which implies in turn that a_{n1} is a free entry, which establishes the lemma.

By Lemma 7.1, if A has signed determinant under (W), there exists a permutation matrix P and a diagonal matrix D where $d_{ii} = +1$ or -1 for every i such that PA^*D has negative diagonal entries in A and satisfies $a_{0j} > 0$ $j = 1,\dots,n$. In the remaining lemmas concerning signed determinants and inverses under (W) given the strong dependence condition, we will assume this "standard" form for A^*.

LEMMA 7.2. *Let $A^* = [a_{ij}]i,\ j = 0,\ldots,n$ satisfy $a_{ij} \neq 0\ i,\ j = 0,\ldots,n$, $a_{jj} < 0 a_{0j} > 0\ j = 1,\ldots,n$. If A has a signed determinant under (W) then $a_{ij} < 0\ i \neq j$ (a_{ij} an entry in A) implies that every cycle $a_{ij}a(j \to i)$ in A involving the entry a_{ij} is negative.*

PROOF. Without loss of generality, assume $a_{12} < 0$ and $a_{12}a_{23}\cdots a_{r-1,r}a_{r1} > 0$. Let $B \in Q_A$ be chosen so that $b_{ii} = -1$, $i = 1,\ldots,n$, $|b_{ij}| = \delta$ for $i \neq j\ i,j = 1,\ldots,n$, where δ is an arbitrarily small positive number. Then $b_{0j} > 0$ can be chosen positive for $j = 1,\ldots,n$, with $\Sigma_{i=0}^{n} b_{ij} = 0\ \ j = 1,\ldots,n$, so that $B^* \in V_{A^*}$, with $\operatorname{sgn}\det B = (-1)^n$. Choose $C \in Q_A$ such that $c_{ii} = -1$, $i = 1,\ldots,n$; $c_{12} = -M$; and $|c_{23}| = |c_{34}| = \cdots = |c_{r-1,r}| = |c_{r1}| = 1 - \delta$, where M and δ are positive numbers. Let $|c_{ij}| = \delta/n$ for every other entry in C. Then $\Sigma_{i=0}^{n} c_{ij} = 0$ implies $c_{02} \geq M + 1 - (n-2)/n\delta$; $c_{0j} \geq 2\delta/n$; $j = 1,3,\ldots,r$; $c_{0j} \geq \delta/n$; and $j = r+1,\ldots,n$, so that for M sufficiently large relative to δ, $C^* \in V_{A^*}$. The positive cycle enters into $\det C$ with sign $(-1)^{n+1}$, and for M sufficiently large, it dominates all other terms in the expansion of $\det C$. Thus $\operatorname{sgn}\det C = (-1)^{n+1}$, while $\operatorname{sgn}\det B = (-1)^n$. Hence A does not have signed determinant under (W).

Intuitively, the reason that negative entries in A must appear only in nonpositive cycles if A is to have a signed determinant under (W) is that given $a_{0j} > 0\ j = 1,\ldots,n$, negative entries in A are free entries. Positive cycles enter $\det A$ with sign opposite to that of the product of negative diagonal entries. Thus if a negative entry appeared in a positive cycle, that positive cycle could be chosen arbitrarily large in value for a matrix $B^* \in V_{A^*}$, and hence A would not have signed determinant under (W).

LEMMA 7.3. *Let A^* satisfy $a_{ij} \neq 0\ i,j = 0,\ldots,n$, where $n \neq 3$, $a_{jj} < 0$, $a_{0j} > 0\ j = 1,\ldots,n$. If A has signed determinant under (W), then at most one column of A has an off-diagonal entry that is negative.*

PROOF. Without loss of generality, assume $a_{21} < 0$ and $a_{rs} < 0$ for $r \neq s$, $s \neq 1$. Assume $r = 1$; then $s \neq 2$, since by Lemma 7.2, $a_{12} < 0 => a_{12}a_{21} < 0$. For $s \neq 2$, consider the cycle $a_{21}a_{1s}a_{s2} < 0$ (by Lemma 7.2), which implies $a_{s2} < 0$, $a_{2s} > 0$. Consider the following cycles of lengths 3 and 4 involving the indices 1, 2, s:

(1) $a_{21}a_{1k}a_{k2} < 0$ implies $a_{1k}a_{k2} > 0$
(2) $a_{s2}a_{2k}a_{ks} < 0$ implies $a_{2k}a_{ks} > 0$
(3) $a_{1s}a_{sk}a_{k1} < 0$ implies $a_{sk}a_{k1} > 0$

(4) $a_{1s}a_{sk}a_{k2}a_{21} < 0$ implies $a_{sk}a_{k2} < 0$
(5) $a_{21}a_{1k}a_{ks}a_{s2} < 0$ implies $a_{1k}a_{ks} < 0$
(6) $a_{s2}a_{2k}a_{k1}a_{1s} < 0$ implies $a_{2k}a_{k1} < 0$

From cycle (6), we have $a_{k1} < 0$ or $a_{2k} < 0$. Suppose $a_{k1} < 0$. Then by Lemma 7.2, $a_{1k} > 0$, while from cycle (6), $a_{2k} > 0$. By cycle (1) this implies in turn that $a_{k2} > 0$. Now cycle (2) implies that $a_{ks} > 0$, while cycle (5) implies that $a_{ks} < 0$. A similar contradiction occurs if $a_{2k} < 0$.

If $r \neq 1$, consider the four cycle $a_{21}a_{1r}a_{rs}a_{s2} < 0$, which implies $a_{1r}a_{s2} < 0$. The above argument can be applied to the three and four cycles involving 1, r, s, 2.

Note that Lemma 7.3 is trivially true for $n = 1$, while Lemma 7.2 excludes the case $a_{12} < 0$, $a_{21} < 0$ for $n = 2$. However, for the case $n = 3$, the conclusion of the lemma does not hold. Thus if A has sign pattern $\begin{bmatrix} - & - & + \\ + & - & - \\ - & + & - \end{bmatrix}$ (with column sums negative, i.e., with $a_{0j} > 0$ $j = 1, 2, 3$), then it can be verified that $C^* \in V_{A^*}$ implies $\det C < 0$ even though every column of A has an off-diagonal negative entry.

LEMMA 7.4. *Let A^* satisfy $a_{ij} \neq 0 i$, $j = 0, \ldots, n$ $(n \neq 3)$, $a_{jj} < 0$, $a_{0j} > 0$ $j = 1, \ldots, n$. If A has signed determinant under (W) and if a column of A has an off-diagonal negative entry, then every off-diagonal entry in that column is negative.*

PROOF. Without loss of generality, assume $a_{21} < 0$, $a_{31} > 0$. By Lemma 7.3, every off-diagonal entry in columns $2, \ldots, n$ is positive. Choose $C \in Q_A$ such that $c_{ii} = -1$ for $i = 1, \ldots, n$, with off-diagonal entries in C arbitrarily small in absolute value, so that $c_{0j} > 0$ $j = 1, \ldots, n$, $C^* \in V_{A^*}$, and $\operatorname{sgn} \det C = (-1)^n$. Choose $B \in Q_A$ such that $b_{11} = -\delta$, $b_{21} = -(1 + \delta)$, $b_{31} = 1$, $|b_{i1}| = \delta/n$ for $i = 4, \ldots, n$, $b_{13} = 1 - \delta$; $b_{ii} = -1$ $i = 2, \ldots, n$, and let $|b_{ij}| = \delta/n$ for all other off-diagonal entries in B. For $\delta > 0$ sufficiently small, $b_{0j} > 0$ $j = 1, \ldots, n$ so that $B^* \in V_{A^*}$. Further, for δ sufficiently small, the product of $b_{13}b_{31}$ with diagonal entries dominates all other terms in the expansion of $|B|$, with $\operatorname{sgn} \det B = (-1)^{n+1}$. Hence A does not have signed determinant under (W).

We next show that if A has negative free diagonal entries and one column of A contains all negative off-diagonal entries, then A has signed determinant under (W) and certain other features of A^{-1} can be identified. The proof utilizes the Perron-Frobenius theorem discussed in the appendix to Chapter 5. For convenience, we restate without proof the basic dominant diagonal characterization of stable Metzler matrices.

LEMMA 7.5 (McKenzie 1960a). *Let C be an $n \times n$ Metzler matrix. Then C is a stable matrix (real parts of the eigenvalues of C are negative) if and only if there exist positive numbers $d_1, \ldots, d_n$ such that*

$$d_i|c_{ii}| > \sum_{j \neq i} d_j|c_{ij}| i = 1, \ldots, n$$

or there exist positive numbers $r_1, \ldots, r_n$ such that

$$r_i|c_{ii}| > \sum_{j \neq i} r_j|c_{ji}| i = 1, \ldots, n.$$

Further, if C is a stable matrix, then $\operatorname{sgn} \det C = (-1)^n$ *and* $\operatorname{sgn} C_{ij} = (-1)^{n-1}$ *for every i, j, where C_{ij} is the cofactor of c_{ij} in C.*

LEMMA 7.6. *Let $A^* = [a_{ij}]i,\ j = 0, \ldots, n$ satisfy $a_{jj} < 0\ j = 1, \ldots, n$, with $a_{i1} < 0\ i = 2, \ldots, n$; $a_{ij} > 0\ i \neq j\ i = 1, \ldots, n,\ j = 2, \ldots, n$. That is, all off-diagonal entries in column 1 of A are negative, while off-diagonal entries in the other columns of A are positive. Then A has signed determinant under (W), with* $\operatorname{sgn} \det A = (-1)^n$; *diagonal entries in A^{-1} are negative; and off-diagonal entries in the first column of A^{-1} are positive, while off-diagonal entries in the first row of A^{-1} are negative. All other entries in A^{-1} are of ambiguous sign.*

PROOF. The proof is by induction. For $n = 1$ or $n = 2$ the lemma is obvious. For $n = 3$, $\det A = a_{11}(a_{22}a_{33} - a_{23}a_{32}) + a_{12}a_{23}a_{31} + a_{13}a_{32}a_{21} - a_{12}a_{21}a_{33} - a_{13}a_{31}a_{22}$. Except for the term within the parentheses, all terms in the expansion of $\det A$ are negative. Moreover, the conditions $a_{02} > 0$, $a_{03} > 0$, $a_{12} > 0$, $a_{13} > 0$ together with the condition that column sums in A^* are zero imply $|a_{22}| > a_{32}$, $|a_{33}| > a_{23}$. Hence $(a_{22}a_{33} - a_{23}a_{32}) > 0$ so that $\det A < 0$. The other properties of A^{-1} are easily verified.

Assume next that the lemma holds for $n = r$. To show that it holds for $n = r + 1$, note that by Lemma 7.5, the $r \times r$ cofactor A_{11} has sign $(-1)^r$, while $\operatorname{sgn} A_{ii} = (-1)^r\ i = 2, \ldots, r + 1$ under the inductive hypothesis. Consider next any cofactor A_{1j} of an entry $a_{1j}\ j = 2, \ldots, r + 1$ in A. $A_{1j} = (-1)^{1+j}x$ determinant formed by crossing out the first row and jth column of A. Perform the following row interchanges within this determinant: interchange row j with $j - 1$, then j with $j - 2, \ldots, j$ with row 2, involving a total of $j - 2$ interchanges. The resulting determinant is an $r \times r$ array with sign pattern (and column sums) such that under the inductive hypothesis, its sign is $(-1)^r$. Hence $\operatorname{sgn} A_{1j} = (-1)^{1+j} (-1)^{j-2}(-1)^r = (-1)^{r+1}$ so that in $\det A = a_{11}A_{11} +, \sum_{j=2}^{r+1} a_{1j}A_{1j}$, each of these terms has sign $(-1)^{r+1}$ so that $\operatorname{sgn} \det A = (-1)^{r+1}$, diagonal entries in A^{-1} are negative, and off-diagonal entries in the first column of A^{-1} are positive. By a similar use of induction, $\operatorname{sgn} A_{i1} = (-1)^r$, $i = 2, \ldots, n$.

Consider next any cofactor A_{ij} of an entry a_{ij} in A where $i \neq j$, $j = 2, \ldots, n$. In any such cofactor, the first column consists solely of negative entries and the ith column contains only positive entries. It is easy to verify that under the sign conditions A_{ij} is ambiguously signed under (W).

COROLLARY. *Under the conditions of Lemma 7.6, A is sign Hicksian under (W); i.e., every ith order principal minor of B has sign $(-1)^i i = 1, \ldots, n$.*

This follows immediately, because either every principal minor of B is of the Metzler sign pattern and dominant diagonal or the first column has all off-diagonal entries negative, with column sums negative, so again by Lemma 7.6, it has the correct sign, for the Hicksian conditions.

The case covered by Lemma 7.6 is that of a matrix with sign pattern of the form

$$\operatorname{sgn} A = \begin{bmatrix} - & + & + & + \\ - & - & + & + \\ - & + & - & + \\ - & + & + & - \end{bmatrix}$$

(with column sums negative), for which

$$\operatorname{sgn} A^{-1} = \begin{bmatrix} - & - & - & - \\ + & - & ? & ? \\ + & ? & - & ? \\ + & ? & ? & - \end{bmatrix},$$

where the question mark denotes a term of ambiguous sign.

From Lemma 7.6, equivalent conditions for signed determinants and signed inverses under (W) given that $a_{ij} \neq 0$ for $i, j = 0, \ldots, n$ are immediate.

THEOREM 7.1. *Let $A^* = [a_{ij}]$ $i,\ j = 0, \ldots, n$ $(n \neq 3)$ satisfy $a_{ij} \neq 0$ $i,\ j = 0, \ldots, n$. A has signed determinant under (W) if and only if there exists a permutation matrix P and a diagonal (signature) matrix D with $d_{ii} = +1$ or -1 for every i such that in $\hat{A}^* = PA^*D$, $a_{jj} < 0$, $a_{0j} > 0 j =$*

1,..., n and either

(1) all off-diagonal entries in $\hat{A}$ are positive; or
(2) in one column of $\hat{A}$ off diagonal entries are negative, while all other off-diagonal entries in $\hat{A}$ are positive (where $\hat{A}$ is the matrix obtained from $\hat{A}^*$ by deleting row and column 0).

THEOREM 7.2. *Let* $A^* = [a_{ij}] i,\ j = 0, \ldots, n\ (n > 2)$ *satisfy* $a_{ij} \neq 0 i,\ j = 0, \ldots, n$. *A has signed inverse under (W) if and only if there exists a permutation matrix P and a diagonal matrix D such that in* $\hat{A}^* = PA^*D$, $a_{jj} < 0,\ a_{0j} > 0\ j = 1, \ldots, n$ *and* $a_{ij} > 0\ i \neq ji,\ j = 1, \ldots, n$.

Using the preceding theorems and lemmas, including the Perron-Frobenius theorem, we obtain a solution to the qualitative comparative statics problem, as summarized in Theorem 7.3.

THEOREM 7.3. *Let* $A^* = [a_{ij}] i,\ j = 0, \ldots, n$ *with* $a_{ij} \neq 0,\ i,\ j = 0, \ldots, n$ $(n > 3)$, $a_{jj} < 0,\ a_{0j} > 0\ j = 1, \ldots, n$. *Then* $Ax = b$ *is fully sign solvable under (W) if and only if either (i) or (ii) holds:*

(i) *if* $b_j \neq 0$ *for more than one index* $j = 1, \ldots, n$, *then every nonzero entry in b has the same sign and* $a_{ij} > 0$ *for* $i \neq j\ i,\ j = 1, \ldots, n$, *and if* $b_j \geq 0$, *then* $x < 0$; *or*
(ii) *if* $b_k \neq 0$ *for a single index* $k (b_j = 0 j \neq k)$, *then either* $a_{ij} > 0$ *for* $i \neq j\ i, j = 1, \ldots, n$ *or* $a_{ik} < 0\ i = 1, \ldots, n$, *with* $a_{ij} > 0\ i \neq j\ j \neq 1$. *In the first case,* $b_k > 0$ *implies* $x_i < 0$ *for all i, while in the second case,* $b_k > 0$ *implies* $x_k < 0$, *with* $x_i > 0\ i \neq k$.

It is easy to verify the status of partial sign solvability under (W), as indicated in Theorem 7.4.

THEOREM 7.4. *Let* $A^* = [a_{ij}] i,\ j = 0, \ldots, n$, *with* $a_{ij} \neq 0\ i,\ j = 0, \ldots, n$ *with* $n > 3$, $a_{jj} < 0,\ a_{0j} > 0\ j = 1, \ldots, n$. *Then* $Ax = b$ *is partially sign solvable under (W) for* x_k *if and only if either*

(1) $Ax = b$ *is fully sign solvable under (W); or*
(2) b_k *is the only nonzero entry in b, and A has signed determinant under (W); or*
(3) $a_{ik} < 0\ i = 1, \ldots, n$, $a_{ij} > 0\ i \neq j\ j \neq k$, *and every nonzero entry* b_i $i \neq k$ *has the same sign, with* b_k *weakly of opposite sign.*

The following examples indicate the scope and nature of the results derived thus far concerning signed determinants, signed inverses, and full sign solvability under (W) for the case in which A^* is full, i.e., where every entry in A^* is nonzero. With the exception of the special case noted earlier for $n \neq 3$, there are only two cases of interest, namely, those identified in Theorem 7.1.

EXAMPLE 1. The Metzler case.

$$\operatorname{sgn} A^* = \begin{bmatrix} - & + & + & + & + \\ + & - & + & + & + \\ + & + & - & + & + \\ + & + & + & - & + \\ + & + & + & + & - \end{bmatrix},$$

with

$$\operatorname{sgn} A = \begin{bmatrix} - & + & + & + \\ + & - & + & + \\ + & + & - & + \\ + & + & + & - \end{bmatrix}.$$

Under (W), sgn det $A = (-1)^n$ (> 0 for $n = 4$), with

$$\operatorname{sgn} A^{-1} = \begin{bmatrix} - & - & - & - \\ - & - & - & - \\ - & - & - & - \\ - & - & - & - \end{bmatrix}.$$

Hence in $Ax = b$, x is fully sign solvable under (W) if and only if (i) $b_i \geq 0$ for all i; or (ii) $b_i \leq 0$ for all i. In case (i), $x = A^{-1}b$ so that

$$\begin{bmatrix} x_1 \\ x_2 \\ x_3 \\ x_4 \end{bmatrix} = \begin{bmatrix} - & - & - & - \\ - & - & - & - \\ - & - & - & - \\ - & - & - & - \end{bmatrix} \begin{bmatrix} + \\ + \\ + \\ + \end{bmatrix} = \begin{bmatrix} - \\ - \\ - \\ - \end{bmatrix},$$

while in case (ii), $x > 0$ for $b \neq 0$.

EXAMPLE 2. One off-diagonal column in A negative.

$$\operatorname{sgn} A^* = \begin{bmatrix} - & + & + & + & + \\ + & - & + & + & + \\ + & - & - & + & + \\ + & - & + & - & + \\ + & - & + & + & - \end{bmatrix},$$

with

$$\text{sgn}\ A = \begin{bmatrix} - & + & + & + \\ - & - & + & + \\ - & + & - & + \\ - & + & + & - \end{bmatrix}.$$

Under (W), sgn det $A = (-1)^n$ (>0 for $n=4$), with

$$\text{sgn}\ A^{-1} = \begin{bmatrix} - & - & - & - \\ + & - & ? & ? \\ + & ? & - & ? \\ + & ? & ? & - \end{bmatrix}.$$

Hence in $Ax=b$, x is fully sign solvable under (W) with $b \neq 0$ if and only if $b_i = 0$ for $i > 1$. When $b_1 > 0$, $b_2 = b_3 = b_4 = 0$, $x = A^{-1}b$ becomes

$$\begin{bmatrix} x_1 \\ x_2 \\ x_3 \\ x_4 \end{bmatrix} = \begin{bmatrix} - & - & - & - \\ + & - & ? & ? \\ + & ? & - & ? \\ + & ? & ? & - \end{bmatrix} \begin{bmatrix} + \\ 0 \\ 0 \\ 0 \end{bmatrix} = \begin{bmatrix} - \\ + \\ + \\ + \end{bmatrix}.$$

Note that if $b_2 > 0$, $b_1 = b_3 = b_4 = 0$, then $x = A^{-1}b$ becomes

$$\begin{bmatrix} x_1 \\ x_2 \\ x_3 \\ x_4 \end{bmatrix} = \begin{bmatrix} - & - & - & - \\ + & - & ? & ? \\ + & ? & - & ? \\ + & ? & ? & - \end{bmatrix} \begin{bmatrix} 0 \\ + \\ 0 \\ 0 \end{bmatrix} = \begin{bmatrix} - \\ - \\ ? \\ ? \end{bmatrix},$$

so that $Ax=b$ is partially sign solvable for x_1 and x_2.

EXAMPLE 3. Special case: $n=2$.

$$\text{sgn}\ A^* = \begin{bmatrix} - & + & + \\ + & - & + \\ + & - & - \end{bmatrix},$$

with

$$\text{sgn}\ A = \begin{bmatrix} - & + \\ - & - \end{bmatrix},$$

so that $\det A > 0$, and

$$\text{sgn } A^{-1} = \begin{bmatrix} - & - \\ + & - \end{bmatrix}.$$

$Ax = b = > x = A^{-1}b$ is fully sign solvable for $b \neq 0$ if and only if $b_1 = 0$ or $b_2 = 0$. For $b_2 = 0$, we have

$$\begin{bmatrix} x_1 \\ x_2 \end{bmatrix} = \begin{bmatrix} - & - \\ + & - \end{bmatrix} \begin{bmatrix} + \\ 0 \end{bmatrix} = \begin{bmatrix} - \\ + \end{bmatrix}.$$

EXAMPLE 4. Special case: $n = 3$.

$$\text{sgn } A^* = \begin{bmatrix} - & + & + & + \\ + & - & - & + \\ + & + & - & - \\ + & - & + & - \end{bmatrix},$$

with

$$\text{sgn } A = \begin{bmatrix} - & - & + \\ + & - & - \\ - & + & - \end{bmatrix}.$$

Under (W) (see the comment earlier), $\det A < 0$, with

$$\text{sgn } A^{-1} = \begin{bmatrix} - & ? & ? \\ ? & - & ? \\ ? & ? & - \end{bmatrix}.$$

Thus, for $b \neq 0$, $Ax = b$ is not fully sign solvable. However, if b_k is the only nonzero entry in b, $Ax = b$ is partially sign solvable under (W) for x_k, $k = 1, 2, 3$.

In the context of the competitive model the comparative statics problem arises because of a change in an environmental parameter, β, in the system

$$\sum_{i=1}^{n} \frac{\partial E_j}{\partial p_i} \frac{dp_i}{d\beta} = -\frac{\partial E_j}{\partial \beta} j = 1, \ldots, n.$$

Hence if all goods are gross substitutes and excess demand functions obey the law of demand (an increase in own price leads to a decrease in excess demand), a shift in the environmental parameter that increases excess demand for one or more commodities leads to an increase in all equilibrium prices, by applying condition (i) of Theorem 7.3. When the kth column of A contains only negative entries, an increase in excess demand for the kth commodity leads to an increase in the equilibrium price of the kth commodity and a fall in all other equilibrium prices, by applying condition (ii) of Theorem 7.3.

A^* Matrices Containing Zeros

In this section we relax the conditions $a_{ij} \neq 0$, $i, j = 0, \ldots, n$ and again consider the implications of Walras's law for qualitatively specified systems. The following condition (condition Q) plays a central role in analysis of such systems (see Habibagahi and Quirk 1972).

Condition Q. Given $a_{rs} \neq 0$, $r \neq s$, and a_{rs} an entry of A, then every positive cycle in A containing a_{rs} also contains all indices k for which $a_{ks} a_{rs} < 0$, where $k \in \{0, 1, \ldots, n\}$; or, alternatively, if I is the index set of a positive cycle in A, then $a_{ij} \geq 0$ for $i \in I'$, $j \in I$, where I' is the complement of I in the set $\{0, 1, \ldots, n\}$.

LEMMA 7.7. *Let $A^* = [a_{ij}]i,\ j = 0, \ldots, n$ satisfy $a_{jj} < 0$ and $a_{0j} > 0$ $j = 1, \ldots, n$. Then A has signed determinant under (W) only if A^* satisfies condition Q.*

PROOF. Without loss of generality, let $a_{12} \neq 0$ and let $a_{12} a_{23} \cdots a_{m1} > 0$ be a positive cycle containing a_{12}. Assume that $a_{k2} a_{12} < 0$, where $k \in \{m+1, \ldots, n\}$. Choose B^* such that $\operatorname{sgn} b_{ij} = \operatorname{sgn} a_{ij}$ $i, j = 0, \ldots, n$, with $b_{ii} = -1$ $i = 1, \ldots, n$, $|b_{23}| = \cdots = |b_{m1}| = 1 - \delta$, $0 < \delta < 1$, and let $|b_{12}| = |b_{k2}| = M > 0$. Choose all other entries b_{ij} in B to satisfy $|b_{ij}| \leq \delta/n$. Set $b_{0j} = -\sum_{i=1}^{n} b_{ij}$. Then $b_{0j} > 0$ for $j = 1, \ldots, n$; hence B^* has the same sign pattern as A^* and has column sums zero so that $B^* \in V_{A^*}$. For M sufficiently large (and δ sufficiently small), the positive cycle dominates all other terms in the expansion of the principal minor of A with indices $(1, 2, \ldots, m)$; hence the sign of that principal minor is $(-1)^{m+1}$. For small values of δ, the sign of $\det B$ is determined by the product of this principal minor with diagonal entries $b_{m+1,m+1}, \ldots, b_{nn}$; hence $\operatorname{sgn} \det B = (-1)^{n+1}$.

Clearly choosing C^* such that $c_{ii} = -1$ $i = 1, \ldots, n$, $|c_{ij}| \leq \varepsilon$ $i \neq j$, $i, j = 1, \ldots, n$, with $c_{0j} = -\sum_{i=1}^{n} c_{ij}$, where $\varepsilon > 0$ is sufficiently small, guarantees that $C^* \in V_{A^*}$, with $\operatorname{sgn} \det C = (-1)^n$. If $k = 0$, so that $a_{02} a_{12} < 0$, then a_{12} is a free entry in A, and again the positive cycle involving a_{12} can be chosen so as to dominate the expansion of $\det B$ so that A does not have signed determinant under (W), which establishes the lemma.

As noted earlier, we say that A is sign Hicksian under (W) if $B^* \in V_{A^*}$ implies that every ith-order principal minor of B has sign $(-1)^i$ $i = 1, \ldots, n$. Then the following is immediate from Lemma 7.7.

COROLLARY. *Under the conditions of Lemma 7.7, A is sign Hicksian under (W) only if A^* satisfies condition Q.*

In *Value and Capital*, Hicks claimed that "extreme complementarity" was a source of instability of the competitive model. We will return to this issue in more detail below, but at this point we note a basic limitation on complementarity ($a_{rs} < 0$ $r \neq s$) imposed by condition Q.

LEMMA 7.8. *Let $A^* = [a_{ij}]i,\ j = 0, \ldots, n$ satisfy $a_{jj} < 0$ and $a_{0j} > 0$ $j = 1, \ldots, n$. If A^* satisfies condition Q, then $a_{rs} < 0$ $r \neq s$ an entry of A implies that every nonzero cycle in A containing a_{rs} is negative.*

PROOF. By condition Q, if $a_{rs} < 0$, $r \neq s$, and a_{rs} appears in a positive cycle in A, then this positive cycle must contain the index 0 since $a_{0j} > 0$ by hypothesis, which provides the desired contradiction.

Condition Q together with Lemma 7.8 thus imply that positive cycles in A contain only positive entries. Further A is sign Hicksian under (W) if and only if it satisfies condition Q.

THEOREM 7.5. *Let $A^* = [a_{ij}]$ $i,\ j = 0, \ldots, n$, satisfy $a_{jj} < 0$ and $a_{0j} > 0$, $j = 1, \ldots, n$. Then A is sign Hicksian under (W) if and only if*

(i) *$a_{ij} < 0$ implies that every nonzero cycle containing a_{ij} is negative; and*
(ii) *A^* satisfies condition Q.*

PROOF. Necessity follows from the corollary to Lemma 7.7. To establish sufficiency, note first that any kth-order principal minor of A containing only nonpositive cycles has sign $(-1)^k$ by the basic Maybee determinant formula. If the principal minor has only nonnegative cycles and there is a nonzero cycle of maximal length (of length k), then by Lemma 7.8, every entry in that cycle must be positive and no off-diagonal entries in the principal minor are negative because if a negative entry existed, it would appear in a nonzero cycle involving some of the entries of the maximal cycle, which would violate either Lemma 7.8 or the assumption that the principal minor contains only nonnegative cycles). By condition Q, column sums of the principal minor must be negative, since there cannot be a negative off-diagonal entry outside the principal minor, in the columns appearing in the principal minor. This implies that the principal submatrix is dominant diagonal, which in turn implies that the principal minor has sign $(-1)^k$. If there is no maximal cycle in the principal minor and $a_{ij} < 0$ for an index j appearing in the principal minor and an index i not in the principal minor, then no entry in the jth column of the principal submatrix appears in a positive cycle in that submatrix. Hence if $a_{ij} < 0$ for some such column indices j, then the principal submatrix is reducible, with any positive cycles appearing only in a dominant diagonal submatrix, while the remaining submatrix has a determinant equal to the product of its (negative) diagonal entries, and once again the sign of the principal minor is $(-1)^k$.

There remains the case where the principal submatrix contains both positive and negative cycles. Consider the case of an $s \times s$ submatrix indexed from 1 to s containing both positive and negative off-diagonal entries in column 1. Let a_{21} be a positive entry appearing in a positive cycle $a_{21}a(1 \to 2)$ and let a_{31} be a negative entry. By condition Q, the positive cycle contains the index 3, so it can be written as $a_{21}a(1 \to 3)a(3 \to 2)$. Since all the entries in the positive cycle are positive, $a_{31}a(1 \to 3)$ is a negative cycle. If these are the only nonzero cycles in the principal minor P,

$$P = a_{11} \cdots a_{ss} - a_{21}a(1 \to 3)a(3 \to 2)\prod_I a_{ii} + a_{31}a(1 \to 3)\prod_J a_{ii}\prod_I a_{ii},$$

where I is the index set of indices $1,\ldots,s$ not appearing in the positive cycle and J is the index set of indices appearing in the positive cycle but not in the negative cycle. Without loss of generality, assume s is even so that the first and third terms in P are positive, while the second term is negative. By (W), $a_{21} < |a_{11}| + |a_{31}|$, since a_{31} is the only negative off-diagonal entry in column 1. Further, the sum of all off-diagonal entries in columns $2,\ldots,s$ is less than the absolute value of the diagonal in that column so that

$$a_{21}a(1 \to 3)a(3 \to 2)\prod_I a_{ii} + a_{31}a(1 \to 3)\prod_{J+I} a_{ii} < (-a_{21} + a_{31})\prod_{i \neq 1} a_{ii}.$$

Hence,

$$P > (a_{11} + a_{31} - a_{21}) \prod_{i \neq I', i \in s} a_{ii} > 0.$$

The argument immediately extends to the case of several positive and negative cycles, given condition Q. Thus if $a_{21} \ldots a_{r1}$ are positive, while $a_{r+1,1} \cdots a_{s1}$ are negative entries in column 1 of A, then any positive cycle $a_{i1}a(1 \to i)$ $i = 2,\ldots,r$ can be rewritten as $a_{i1}a(1 \to k)$ $a(k \to i)$ for $k = r+1,\ldots,s$, so that, by the restrictions imposed by (W), there are terms in the expansion of P, $a_{i1}a(1 \to k)\ a(k \to i) - a_{ki}a(1 \to k)\ \prod a_{ii}$, such that $|a(1 \to k)\prod a_{ii}| > |a(1 \to k)a(k \to i)|$, which in turn, when combined with the condition $|a_{11} + \Sigma_{i=r+1}^{s} a_{i1}| > \Sigma_{i=2}^{r} a_{i1}$, implies that P is signed $(-1)^s$.

THEOREM 7.6. *Under the conditions of Theorem 7.5, A has signed determinant under (W) if and only if*

(i) $a_{ij} < 0$ *implies that every nonzero cycle containing* a_{ij} *is negative; and*
(ii) A^* *satisfies condition Q.*

PROOF. This follows immediately from Lemma 7.7 and Theorem 7.5.

The following illustrates the application of Theorem 7.6:

$$\text{sgn } A^* = \begin{bmatrix} - & + & + & + & + \\ + & - & + & + & 0 \\ + & + & - & + & 0 \\ + & + & + & - & - \\ + & + & + & + & - \end{bmatrix},$$

which, by inspection, satisfies condition Q. Sgn$|A|$ can be determined by writing out its expansion explicitly:

$$|A| = a_{44}\begin{vmatrix} a_{11} & a_{12} & a_{13} \\ a_{21} & a_{22} & a_{23} \\ a_{31} & a_{32} & a_{33} \end{vmatrix} - a_{34}\begin{vmatrix} a_{11} & a_{12} & a_{13} \\ a_{21} & a_{22} & a_{23} \\ a_{41} & a_{42} & a_{43} \end{vmatrix}.$$

Note that the term multiplying a_{44} is negative because it is the determinant of a 3×3 dominant diagonal matrix. Writing out the second term, we have

$$(-)\{(-)(+)((+)) + (-)(+)((+)(+) - (+)(-))$$
$$-(-)(+)((-)(+) - (+)(+))\}$$

$$(-)\{a_{34}a_{43}(a_{11}a_{22} - a_{12}a_{21})$$
$$+a_{34}a_{13}(a_{21}a_{42} - a_{41}a_{22}) - a_{34}a_{23}(a_{11}a_{42} - a_{12}a_{41})\}.$$

Every term inside the braces is negative; hence sgn$|A| = (+)$. (The expression $(a_{11}a_{22} - a_{12}a_{21})$ is positive, again by the dominant diagonal argument.)

Given Theorem 7.5 and the Maybee determinant formula (see the appendix to Chapter 2), the following result is immediate.

LEMMA 7.9. *Let* $A^* = [a_{ij}]i,\ j = 0,\dots,n$ *satisfy* $a_{jj} < 0$ *and* $a_{0j} > 0 j = 1,\dots,n$. *Then* A_{ij} $i \neq j$ *is signed under* (W) *if and only if*

(i) $a_{ij} < 0$ *implies that every nonzero cycle containing* a_{ij} *is negative;*
(ii) A^* *satisfies condition* Q; *and*
(iii) *every nonzero path* $a(j \to i)$ *in* A *has the same sign.*

THEOREM 7.7. *Let* $A^* = [a_{ij}]\ i,\ j = 0,\dots,n$ *satisfy* $a_{jj} < 0$ *and* $a_{0j} > 0$ $j = 1,\dots,n$. *Then* A *has signed inverse under* (W) *if and only if*

(i) $a_{ij} < 0$ *implies that every nonzero cycle containing* a_{ij} *is negative;*
(ii) A^* *satisfies condition* Q; *and*
(iii) *for every* $i \neq j$ $\quad i, j = 1,\dots,n$, *every nonzero path* $a(j \to i)$ *in* A *has the same sign.*

These results lead into the following results concerning sign solvability under (W).

THEOREM 7.8. *Let* $A^* = [a_{ij}]\ i,\ j = 0,\dots,n$ *satisfy* $a_{jj} < 0,\ a_{0j} > 0\ j = 1,\dots,n$. *Then* $Ax = b$ *is fully sign solvable under* (W) *if and only if*

(i) $a_{ij} < 0$ *implies that every nonzero cycle containing* a_{ij} *is negative;*
(ii) A^* *satisfies condition* Q; *and*
(iii) $b_k \neq 0$ *implies that for any* $i = 1,\dots,n$, *all nonzero paths* $a(i \to k)$ *have the same sign and that every nonzero term* $b_k A_{ki}$ *has the same sign for a given* i.

THEOREM 7.9. *Let* $A^* = [a_{ij}]i,\ j = 0,\dots,n$ *satisfy* $a_{jj} < 0,\ a_{0j} > 0\ j = 1,\dots,n$. *Then* $Ax = b$ *is partially sign solvable under* (W) *for* x_j *if and only if*

(i) $a_{ij} < 0$ *implies that every nonzero cycle containing* a_{ij} *is negative;*
(ii) A^* *satisfies conditions* Q; *and*
(iii) $b_k \neq 0$ *implies that all nonzero paths* $a(j \to k)$ *have the same sign and that every nonzero term* $b_k A_{kj}$ *has the same sign.*

These results can be illustrated by the following examples:

$$\operatorname{sgn} A^* = \begin{bmatrix} - & + & + & + & + \\ + & - & + & + & 0 \\ + & + & - & + & 0 \\ + & + & + & - & - \\ + & + & + & + & - \end{bmatrix},$$

and

$$\operatorname{sgn} A = \begin{bmatrix} - & + & + & 0 \\ + & - & + & 0 \\ + & + & - & - \\ + & + & + & - \end{bmatrix}.$$

A^* satisfies condition Q so that A is sign Hicksian, with

$$\operatorname{sgn} A^{-1} = \begin{bmatrix} - & ? & - & + \\ ? & - & - & + \\ ? & ? & - & + \\ + & - & + & - \end{bmatrix}.$$

Thus if

$$b = \begin{bmatrix} 0 \\ 0 \\ + \\ - \end{bmatrix},$$

$Ax = b$ implies

$$\begin{bmatrix} x_2 \\ x_2 \\ x_3 \\ x_4 \end{bmatrix} = \begin{bmatrix} - & ? & - & + \\ ? & - & - & + \\ ? & ? & - & - \\ + & - & + & - \end{bmatrix} \begin{bmatrix} 0 \\ 0 \\ + \\ - \end{bmatrix} = \begin{bmatrix} - \\ - \\ - \\ + \end{bmatrix},$$

and there is full sign solvability under (W). If

$$b = \begin{bmatrix} + \\ - \\ + \\ - \end{bmatrix},$$

with A^* as above, then

$$\begin{bmatrix} x_1 \\ x_2 \\ x_3 \\ x_4 \end{bmatrix} = \begin{bmatrix} - & ? & - & + \\ ? & - & - & + \\ ? & ? & - & - \\ + & - & + & - \end{bmatrix} \begin{bmatrix} + \\ - \\ + \\ - \end{bmatrix} = \begin{bmatrix} ? \\ ? \\ ? \\ + \end{bmatrix},$$

and $Ax = b$ is partially sign solvable under (W) for x_4.

7.6 SIGN SOLVABILITY UNDER WALRAS'S LAW AND HOMOGENEITY

The more interesting case of sign solvability, from an economist's point of view, is that in which both (W) and (H) hold. As earlier, let

$$S_{A^*} = \left\{ B^* \in Q_{A^*} \text{ such that } B^* \text{ satisfies (W) } \sum_{j=0}^{n} b_{ij} = 0, i = 0, \ldots, n \right.$$

$$\left. \text{and (H) } \sum_{i=0}^{n} b_{ij} = 0, j = 0, \ldots, n \right\};$$

then we say that $Ax = b$ is *fully sign solvable under* (W) *and* (H) if $B^* \in S_{A^*}$, $c \in Q_b$, $By = c$ implies that $y \in Q_x$; and we say that $Ax = b$ is *partially sign solvable under* (W) *and* (H) if some subset of x is sign solvable under (W) and (H).

In investigating sign solvability under (W) and (H), the concept of a free entry is again important; a_{rs} is said to be a free entry in A under (W) and (H) if given any real number $M > 0$, there exists $C^* \in S_{A^*}$ such that $|c_{rs}| > M|c_{ij}|$ for every $i \neq r$ or $j \neq s$, $i, j = 1, \ldots, n$.

It is clear that a_{rs} is a free entry in A under (W) and (H) if and only if $a_{rs}a_{r0} < 0$, $a_{rs}a_{0s} < 0$, and $a_{rs}a_{00} > 0$. Further, if A has signed determinant under (W) and (H), it is the case that if a_{rs} is a free entry in A under (W) and (H) and $A_{rs} \neq 0$, where A_{rs} is the cofactor of a_{rs} in A, then $\operatorname{sgn} \det A = \operatorname{sgn} a_{rs}A_{rs}$. Similarly, if A has signed determinant under (W) and (H), then $\det A$ has the same sign as any nonzero term in its expansion involving $n-1$ free entries under (W) and (H).

An interesting property of the matrix A^* under (W) and (H) is the following.

THEOREM 7.10. *If A^* satisfies (W) and (H), then every $n \times n$ cofactor A^*_{ij} of A^* is equal.*

PROOF. If $A^*_{ij} = 0$ for all $i, j = 0, \ldots, n$, then the theorem holds. Hence assume that $A^*_{ij} \neq 0$ for some $i, j = 0, \ldots, n$. Without loss of generality, reindex so that $A^*_{00} \neq 0$. By (W) and (H) we have

$$\sum_{j=0}^{n} a_{ij} = 0, i = 0, \ldots, n, \quad \sum_{i=0}^{n} a_{ij} = 0, j = 0, \ldots, n.$$

Further, $\Sigma_{j=0}^{n} a_{ij} A_{ij}^* = 0$, $i = 0, \ldots, n$, $\Sigma_{i=0}^{n} a_{ij} A_{ij}^* = 0$, $j = 0, \ldots, n$, since A^* is a singular matrix (in particular, $A^* d^* = 0$, where d^* is a vector with $d_i = 1$, $i = 0, \ldots, n$). Thus,

$$\sum_{j=1}^{n} a_{ij} A_{ij}^* = -a_{i0} A_{i0}^*, i = 1, \ldots, n, \text{ and } \sum_{j=1}^{n} a_{ij} = -a_{i0}, i = 1, \ldots, n.$$

Hence, $\Sigma_{j=1}^{n} a_{ij}(A_{ij}^* - A_{i0}^*) = 0$, $i = 1, \ldots, n$. But $[a_{ij}]$, $i, j = 1, \ldots, n$ has determinant $A_{00}^* \neq 0$ by hypothesis; hence $A_{ij}^* = A_{i0}^*$, $i = 1, \ldots, n$, and similarly, $A_{ij}^* = A_{0j}^*$, $j = 1, \ldots, n$. Also, $\Sigma_{j=1}^{n} a_{0j} A_{0j}^* = -a_{00} A_{00}^*$ and $\Sigma_{j=1}^{n} a_{0j} = -a_{00}$. Since $A_{0j}^* = A_{i0}^*$, $i, j = 1, \ldots, n$, thus $-a_{00} A_{00}^* = A_{i0}^* \Sigma_{j=1}^{n} a_{0j}$ implies $A_{00}^* = A_{i0}^*$, $i = 1, \ldots, n$. Hence, $A_{ij}^* = A_{i0}^* = A_{0j}^* = A_{00}^*$, $i, j = 1, \ldots, n$.

As in the preceding section, we will deal only with the case in which the law of demand holds for all goods, and the *numeraire* is a substitute for all other goods and conversely, i.e., $a_{ii} < 0$, $i = 0, \ldots, n$, $a_{i0} > 0$, $i = 1, \ldots, n$, $a_{0j} > j = 1, \ldots, n$. The arguments to be presented can be extended to other cases, but only at the cost of a much more complicated notation. The case under consideration is the only qualitatively specified case that has received any extensive treatment in the existing literature. In this case, negative entries in A, including the diagonal entries, are free entries in A under (W) and (H).

LEMMA 7.10. *Assume that the numeraire is a substitute for all other goods and conversely, and that the law of demand holds for all goods. If A has signed determinant under (W) and (H), then $a_{ij} < 0$, $i \neq j$ implies that a_{ij} appears only in nonpositive cycles.*

PROOF. Assume there exists a positive cycle of length r in A involving $a_{ij} < 0$. Without loss of generality, reindex the cycle into $a_{12} a_{23} \ldots a_{r-1,r} a_{r1}$, where $a_{12} < 0$. Let $B^* \in Q_{A^*}$ be chosen so that $b_{ii} = -1$, $i = 1, \ldots, n$, $|b_{ij}| = \delta$ for nonzero entries b_{ij}, $i \neq j$, $i, j = 1, \ldots, n$, where δ is an arbitrarily small positive number. Then,

$$b_{i0} = -\sum_{j=1}^{n} b_{ij} > 0, b_{0j} = -\sum_{i=1}^{n} b_{ij} > 0, i, j = 1, \ldots, n \text{ and } b_{00}$$
$$= -\sum_{i=1}^{n} b_{i0} < 0;$$

hence $B^* \in S_{A^*}$, with $\operatorname{sgn}|B| = (-1)^n$.

Choose $C^* \in Q_{A^*}$, with $c_{12} = -M$, $|c_{23}| = \cdots = |c_{r1}| = 1$, $c_{ii} = -2$, $i = 1,\ldots,n$, $|c_{ij}| = \delta$ for every other nonzero off-diagonal entry in A where M is positive. For δ sufficiently small, $c_{i0} > 0$, $i = 1,\ldots,n$, $c_{0j} > 0$, $j = 1,\ldots,n$ and $c_{00} < 0$ so that $C^* \in S_{A^*}$. Further, $\det C = (-2)^n + (-1)^{r+1}(-2)^{n-r}M +$ terms arbitrarily small in absolute value. For an appropriate choice of M and δ, $\operatorname{sgn}\det C = (-1)^{n+1}$; hence A does not have signed determinant under (W) and (H).

The *direct* link between market i and market j can be interpreted as $a_{ij} = \partial E_j / \partial p_i$. Paths of the form $a(j \to i)$ with two or more links then can be thought of as measuring the *indirect* link between markets i and j. Lemma 7.10 then asserts that in the case where the law of demand holds and the *numeraire* is a substitute for other goods and conversely, if A has signed determinant under (W) and (H) and j is a direct complement for i ($a_{ji} < 0$), then, directly and indirectly, i must be a substitute for j.

In both Hicks's and Morishima's approaches, sign symmetry of A^* was assumed; i.e., it was assumed that if good i is a substitute for good j, then good j is a substitute for good i, and similarly for complements. Sign symmetry is not implied by the competitive assumptions unless side conditions are imposed, e.g., that income effects are zero, or the Arrow-Hurwicz case of a pure trade economy in which no trades take place at equilibrium. Here, we see that Lemma 7.10 precludes sign symmetry, except in the case where all goods are substitutes for one another. No pair of goods i and j can be direct complements if A has signed determinant under (W) and (H). Moreover, getting ahead of things a bit, stability of A implies that $\operatorname{sgn}\det A = (-1)^n$, so the presence of pairs of complements such as i and j also precludes the proof of (linear approximation) stability simply from sign pattern information together with (W) and (H). While Hicks was concerned, of course, with Hicksian rather than dynamic stability, still this might offer at least a clue as to why Hicks regarded complements as a potential problem.

The following condition, an extension of condition Q above, provides another constraint on the presence of negative off-diagonal entries in A.

Condition Q′. Let I denote the index set of a positive cycle in A. Then either $a_{ij} \geq 0$, $i \in I$, $j \notin I$, or $a_{ij} \geq 0$, $i \notin I$, $j \in I$.

The idea behind condition Q′ is to eliminate the possibility that positive cycles can dominate the expansion of any principal minor in A, since they enter with sign opposite to that of the product of diagonal entries. In effect, condition Q′ ensures that entries of positive cycles in A are not free entries in the principal minor with index set that of the positive cycle. The importance of condition Q′ is indicated in the following lemma.

LEMMA 7.11. *If the numeraire is a substitute for all other goods and conversely, and if the law of demand holds for all goods, then A has signed determinant under (W) and (H) if and only if*

(i) $a_{ij} < 0$ $i \neq j$ *implies that* a_{ij} *appears only in nonpositive cycles in* A; *and*
(ii) *condition* Q' *holds.*

PROOF. The necessity of (i) follows from Lemma 7.10. For the necessity of (ii), given that I is the index set of a positive cycle in A, first assume that there exist $a_{ir} < 0$, $a_{sj} < 0$, for $i, j \in I$, $r, s \notin I$, $i \neq j$. By (i), every entry in any positive cycle in A must be positive. Thus an entry a_{ij} in the positive cycle is a free entry within the principal minor with index set I. That is, $B^* \in S_{A^*}$ can be chosen with b_{ij} arbitrarily large in absolute value relative to all other entries in the principal minor, with appropriate values opposite in sign being assigned to a_{ir} and a_{sj} so as to satisfy row and column sums in B^* equal to zero. Since positive cycles enter into the expansion of an $m \times m$ principal minor with sign $(-1)^{m+1}$, the sign of the principal minor is $(-1)^{m+1}$. Off-diagonal entries in the rest of A can be chosen so that the term dominating $\det A$ is the product of the principal minor with diagonal entries, so that $\operatorname{sgn} \det A = (-1)^{n+1}$. On the other hand, the diagonal entries in A are free entries, so it is also possible to choose them arbitrarily large in absolute value relative to other entries in A, so that the sign of the principal minor is $(-1)^m$ and the sign of the determinant is $(-1)^n$. Hence A does not have signed determinant under (W) and (H).

Given $a_{ir} < 0$, $a_{sj} < 0$, if $a_{ij} \leq 0$, for an entry a_{ij} in the principal minor with index set $\{1, \ldots, m\}$, reindex the principal minor so that the positive cycle can be written as $a_{12} a_{23} \ldots a_{m-1,m} a_{m1}$, and, without loss of generality, assume that $a_{1r} < 0$, $a_{s3} < 0$.

Then $a_{12} < |a_{22}|$, $a_{23} < |a_{22}|$, while $a_{i,i+1} < |a_{ii}|$, $i = 3, \ldots, m-1$, with $a_{m1} < |a_{11}|$. Then there exists $B^* \in S_{A^*}$ such that $b_{ii} = -1$, $i = 1, 3, \ldots, m$, $b_{22} = -M$, with $b_{i,i+1} = 1 - \varepsilon$, $i = 3, \ldots, m-1$, $b_{m1} = 1 - \varepsilon$, with $b_{12} = b_{23} = M - \varepsilon$, $b_{1r} = b_{s3} = -M$, and with all other nonzero entries in B arbitrarily small in absolute value. For M sufficiently large relative to ε, the positive cycle dominates the expansion of the principal minor with index set I, and hence, by the previous argument, A does not have signed determinant under (W) and (H).

Similarly, if $a_{ir} < 0$, $a_{si} < 0$, and a_{ki} is a nonzero entry in the positive cycle, then a_{ki} can be chosen large relative to a_{ii}, without violating the row and column sum restriction, by choosing a_{ir} and a_{si} large in absolute value, as in the earlier example. Once again, the positive cycle dominates the expansion of the principal minor, which in turn means that A does not have signed determinant under (W) and (H).

For sufficiency, first note that, by the Perron-Frobenius theorem, if a matrix is Metzler, with negative diagonal entries and nonnegative off-diagonal entries, then the matrix is stable if and only if it is dominant diagonal and it is dominant diagonal if and only if it is Hicksian. Under condition Q′, if a principal minor with index set I contains only nonnegative cycles and contains a positive cycle with index set I, then it is either row or column dominant diagonal. This is because all the off-diagonal entries outside the principal minor (including the *numeraire* entries) are nonnegative for either all the row indices or all the column indices in I. If the principal minor does not contain a positive cycle of maximal length, still each principal submatrix of the principal minor containing a positive cycle is row or column dominant diagonal. Because of condition Q′, positive cycles in the principal minor are dominated by products of diagonal entries. Thus no entries in positive cycles in the principal minor are free entries within the principal minor. The consequence is that the expansion of the principal minor itself is dominated by the product of diagonal entries, and thus has the Hicksian sign.

Any principal minor of A that contains only nonpositive cycles is signed Hicksian simply because of its sign pattern alone, independently of condition Q′. Finally, consider any principal minor with index set I containing a maximal positive cycle $a_{kj}a(j \to k)$ and a negative entry a_{ij} $i \neq j$. Since the positive cycle is maximal, $a(j \to k)$ includes the index i; hence there exists a negative cycle $a_{ij}a(j \to i)$ containing entries from the positive cycle. Assuming a_{ij} is the only off-diagonal negative entry in column j, we have $a_{kj} < |a_{ji}| + |a_{ij}|$; hence $a_{kj}a(j \to k) < |a_{jj}\Pi_{r \in I/j}\, a_{sr}| + |a_{ij}a(j \to i)\Pi_{r \in k}\, a_{ir}|$, where $\Pi_{r \in I/j}\, a_{rr}$ is the product of diagonal entries with indices in I other than j and $\Pi_{r \in k}\, a_{rr}$ is the product of diagonal entries with indices appearing in $a_{kj}a(j \to k)$ and not in $a_{ij}a(j \to i)$. Thus positive cycles are dominated in the expansion of the principal minor by negative cycles and/or products of diagonal entries, which enter with the Hicksian sign.

Alternatively, note that if negative entries in the principal minor are set equal to zero, condition Q′ ensures that the principal minor is either row or column dominant diagonal, which means it is Hicksian. Increasing the absolute value of any off-diagonal negative entry increases the value of a principal minor of even dimension and decreases the value of a principal minor of odd dimension, so that, again, condition Q′ ensures that the principal minor has the Hicksian sign.

Since det A is the nth-order principal minor of A, this means that (i) and (ii) are sufficient for signing the determinant of A under (W) and (H).

COROLLARY. *If the numeraire is a substitute for all other goods and conversely and the law of demand holds for all goods, then A is sign Hicksian under (W) and (H); i.e., $B^* \in S_{A^*}$ implies that B is Hicksian if and only if (i) and (ii) of Lemma 7.11 hold.*

THEOREM 7.11. *If the numeraire is a substitute for all other goods and conversely and the law of demand holds for all goods, then $Ax = b$ is fully sign solvable under (W) and (H) if and only if*

(i) *A is sign Hicksian under (W) and (H), i.e.,*
  (a) *$a_{ij} < 0$ $i \neq j$ implies that a_{ij} appears only in nonpositive cycles in A, and*
  (b) *condition Q' holds; and*
(ii) *$b_k \neq 0$ implies that every nonzero path $b(k \to i)$ has weakly the same sign for any i, and, for any i, $B_{ik} b_k$ has weakly the same sign for every k.*

PROOF. $B^* \in S_{A^*}$ implies that B^{-1} exists, with diagonal elements in B^{-1} being negative. By the corollary to lemma 7.11, (i) is necessary and sufficient for this. Given (i), the Maybee formula implies that (ii) is sufficient for sign solvability. And, given (i), if (ii) does not hold, clearly there exist C^*, $E^* \in S_{A^*}$ such that sgn $C^*_{ik} \neq$ sgn E^*_{ik}, which for $b_k \neq 0$, implies $Ax = b$ is not sign solvable for some x_i $i = 1, \ldots, n$.

Equivalent conditions for partial sign solvability under (W) and (H) follow immediately from Theorem 7.11.

THEOREM 7.12. *If the numeraire is a substitute for all other goods and conversely, if the law of demand holds for all goods, and if A is irreducible, then $Ax = b$ is partially sign solvable for x_r if and only if A is sign Hicksian under (W) and (H) and $b_k \neq 0$ implies that every chain $b(k \to r)$ has weakly the same sign, with $B_{rk} b_k$ having weakly the same sign for every k.*

To illustrate these theorems, suppose that there is an exogenous increase in excess demand for good 1 in the comparative statics system $x = A^{-1}b$, where x_i measures the change in the ith equilibrium price due to the disturbance. Among the cases for which definite qualitative comparative statics results can be obtained from sign pattern information and (W) and (H) are the case where all goods are gross substitutes (case 1 below); the case where there are only nonpositive products of direct a_{ij} and indirect $a(j \to i)$ links among markets (case 2 below); the case where good 1 is a complement for all other non-*numeraire* goods, which are all substitutes for other goods (case 3 below); and two cases of mixed complementary and substitute relationships among goods (cases 4 and 5 below).

CASE 1. Gross substitute case.

$$A^* = \begin{bmatrix} - & + & + & + \\ + & - & + & + \\ + & + & - & + \\ + & + & + & + \end{bmatrix}, A = \begin{bmatrix} - & + & + \\ + & - & + \\ + & + & - \end{bmatrix},$$

$$x = \begin{bmatrix} - & - & - \\ - & - & - \\ - & - & - \end{bmatrix} \begin{bmatrix} - \\ 0 \\ 0 \end{bmatrix} = \begin{bmatrix} + \\ + \\ + \end{bmatrix}.$$

This illustrates the first two of Hicks's three laws of comparative statics: an exogenous increase in excess demand for a good increases the price of the good and also increases the prices of all other goods in a gross substitute system.

CASE 2. Good i is a complement for $i+1$, and $i+1$ is a substitute for good i, for all i.

$$A^* = \begin{bmatrix} - & + & + & + \\ + & - & + & 0 \\ + & - & - & + \\ + & 0 & - & - \end{bmatrix}, A = \begin{bmatrix} - & + & 0 \\ - & - & + \\ 0 & - & - \end{bmatrix},$$

$$x = \begin{bmatrix} - & - & - \\ + & - & - \\ + & + & - \end{bmatrix} \begin{bmatrix} - \\ 0 \\ 0 \end{bmatrix} = \begin{bmatrix} + \\ - \\ - \end{bmatrix}.$$

This is a case in which A contains only nonpositive cycles, with an exogenous increase in excess demand for good 1 leading to an increase in the price of good 1 and a decrease in the equilibrium prices of the other goods.

CASE 3. Good 1 is a complement for other goods, which are substitutes for all goods.

$$A^* = \begin{bmatrix} - & + & + & + \\ + & - & + & + \\ + & - & - & + \\ + & - & - & - \end{bmatrix}, A = \begin{bmatrix} - & + & + \\ - & - & + \\ - & + & - \end{bmatrix},$$

$$x = \begin{bmatrix} - & - & - \\ + & - & ? \\ + & ? & - \end{bmatrix} \begin{bmatrix} - \\ 0 \\ 0 \end{bmatrix} = \begin{bmatrix} + \\ - \\ - \end{bmatrix}.$$

In this case, there is full sign solvability even though there are unsigned entries in A^{-1} (indicated by question marks). An exogenous increase in excess demand for good 1 increases its price and reduces the prices of the other two non-*numeraire* goods.

CASE 4. Goods 1 and 2 are substitutes for each other, good 3 is a complement for good 2, and good 2 is a substitute for good 3.

$$A^* = \begin{bmatrix} - & + & + & + \\ + & - & + & 0 \\ + & + & - & + \\ + & 0 & - & - \end{bmatrix}, A = \begin{bmatrix} - & + & 0 \\ + & - & + \\ 0 & - & - \end{bmatrix},$$

$$x = \begin{bmatrix} - & - & - \\ - & - & - \\ + & + & - \end{bmatrix} \begin{bmatrix} - \\ 0 \\ 0 \end{bmatrix} = \begin{bmatrix} + \\ + \\ - \end{bmatrix}.$$

This case involves both negative and positive cycles in A, with an increase in excess demand for good 1 leading to an increase in its price and the price of its substitute and a decrease in the price of good 3, which is an independent good relative to good 1, but is a complement for good 2.

CASE 5. Asymmetric complementarity-substitute relations among all goods, a case of partial, but not full, sign solvability.

$$A^* = \begin{bmatrix} - & + & + & + \\ + & - & + & - \\ + & - & - & + \\ + & + & - & - \end{bmatrix} A = \begin{bmatrix} - & + & - \\ - & - & + \\ + & - & - \end{bmatrix}$$

$$x = \begin{bmatrix} - & - & ? \\ ? & - & - \\ - & ? & - \end{bmatrix} \begin{bmatrix} - \\ 0 \\ 0 \end{bmatrix} = \begin{bmatrix} + \\ ? \\ + \end{bmatrix}.$$

In this case, there is a negative cycle of length 3 and a positive cycle of length 3, along with negative cycles of length 2. An increase in excess demand for good 1 leads to an increase in its price and the price of good 3, a substitute for good 1, and an indeterminate effect on the price of good 2, which is a complement for good 1 and a substitute for good 3.

Among the cases not covered by the theorems or the examples above is the Morishima case, where there is sign symmetry of A^*, with "substitutes of substitutes and complements being substitutes, and complements of substitutes and substitutes of complements being complements". As noted earlier, if there is a pair of goods i and j that are direct complements for each other, then A does not have signed determinant under (W) and (H) when the *numeraire* is a substitute for other goods and conversely. The Morishima comparative statics theorems relating to this case hold only when additional quantitative restrictions beyond (W) and (H) are imposed, sufficient to guarantee either a row or column dominant diagonal property.

One important specialization of Theorems 7.11 and 7.12 is the case where A is a "full" matrix, i.e., where all entries in A are nonzero. It is easy to verify in this case that if A has negative off-diagonal entries a_{ij} and a_{rs}, $i \neq r$, $j \neq s$, then either A does not have signed determinant under (W) and (H) or, if a signed determinant exists, there are ambiguously signed entries in every column of the inverse matrix, as in case 5 above.

The following result holds.

THEOREM 7.13. *If the numeraire is a substitute for all other goods and conversely, and if the law of demand holds for all goods, and if* $a_{ij} \neq 0$ $i, j = 1, \ldots, n$, *then* $Ax = b$ *is fully sign solvable under* (W) *and* (H) *if and only if*

(i) *A is a Metzler matrix, with all nonzero entries in b of the same sign; or*

(ii) *off-diagonal entries in the kth row or column of A are negative, with all other off-diagonal entries positive, and b has one nonzero entry, in the kth position; or*

(iii) *there is exactly one negative off-diagonal entry in A, a_{rs}, with all other off-diagonal entries in A being positive, and b has one nonzero entry, in the rth position.*

PROOF. Note that (i) is clearly sufficient for sign solvability, since $B^* \in S_{A^*}$ implies that B is dominant diagonal, with all entries in B^{-1} negative. An instance of (ii) is case 3 above. Under the conditions given, $B^* \in S_{A^*}$ implies that in B^{-1}, the kth row and column are signed, with ambiguously signed entries in every other row and column of B^{-1}, as noted in the earlier discussion. With b_k the only nonzero entry in b, $Ax = b$ is then sign solvable under (ii). If (iii) holds, then all off-diagonal cofactors of entries in row r and column s are signed opposite to that of $\det A$, since the matrix is Hicksian, and all paths appearing in these cofactors are positive. No other off-diagonal cofactors are signed, because the negative entry a_{rs} appears in certain negative paths in these cofactors, which also have positive paths appearing in them. Finally, (iii) is sufficient for sign solvability, because $\operatorname{sgn} A^*_{rr} = (-1)^n$ since it is Metzler and dominant diagonal, while, by Lemma 7.10, all $n \times n$ cofactors in A^* are equal; hence $\operatorname{sgn} \det A = (-1)^n$. Cofactors of entries in the rth row of A are all signed, since A is Hicksian by the corollary to Lemma 7.11 and all paths appearing in these cofactors are positive. Hence (iii) implies sign solvability.

For necessity, as noted above, if negative off-diagonal entries appear in two rows and columns of A, then either A does not have signed determinant under (W) and (H) or ambiguously signed entries appear in every row and column of the inverse. With $n-1$ off-diagonal entries appearing in the kth column of A, if that column contains more than one and less than $n-1$ negative off-diagonal entries appearing in the kth column of A, then there will be paths of opposite sign appearing in off-diagonal cofactors in every column of the inverse matrix. Thus no column of the inverse will be signed; hence $Ax = b$ is not sign solvable.

8

The Competitive Equilibrium: Stability

8.1 INTRODUCTION

This chapter examines the application of qualitative concepts to the problem of determining the stability of a model. As before, our example is the competitive equilibrium. After restating what is meant by stability in this setting, we examine the implications of Walras's law, and homogeneity for identifying qualitatively specified competitive models where stability can be proved.

In our analysis of the stability properties of the competitive model we employ the Walrasian *tâtonnement* mechanism as formalized in Arrow and Hurwicz (1958, 1959). Under this mechanism, it is assumed that a "referee" is charged with the job of clearing markets in the economy. He or she follows the rules of increasing the price of any commodity for which excess demand is positive and reducing the price of any commodity for which excess demand is negative. At a "calling out" of a price vector by the referee, all consumers and firms communicate to the referee the amounts of various commodities they wish to demand or supply, but actual trades take place only at equilibrium; hence the resource endowments remain fixed throughout the adjustment process. If the time path of prices approaches an equilibrium price vector asymptotically, we say that the process is (dynamically) stable.

Formally, the *tâtonnement* mechanism (T) may be expressed as follows:

$$\text{(T)} \quad \dot{p}_j = g_j\left[E_j(p)\right], j = 1, \ldots, n,$$

where $\dot{p}_j = dp_j/dt$ and t denotes time, $g_j(\cdot)$ is an increasing function of $E_j(p)$, and $g_j(0) = 0$. Thus (T) asserts that if excess demand for good j is positive, the price of good j is increased, while if excess demand for good j is negative, the price of good j is decreased. At a competitive equilibrium p^* such that $E_j(p^*) = 0$ for every j, no change in prices takes place. Because good 0 has been chosen as *numeraire*, no change in its price takes place over time.

Let $p^* = (1, p_1^*, \ldots, p_n^*)$ denote a competitive equilibrium price vector. Let $\mathring{p} = (1, \mathring{p}_1, \ldots, \mathring{p}_n)$ denote an arbitrary initial price vector, e.g., the first price vector called out by the referee. Let $f_j(t; p^0)$ denote the time path of p_j as determined by the system (T). Following Arrow and Hurwicz, we distinguish among three notions of stability of (T):

1. p^* is said to be *globally stable* if, given any p^0, $\lim_{t\to\infty} f_j(t; p^0) = p_j^*$, $j = 1, \ldots, n$.
2. p^* is said to be *locally stable* if there exists a neighborhood

$$N_\delta(p^*) = \left\{ p \mid \sum_{j=1}^{n} (p_j - p_j^*)^2 < \delta \right\}$$

   such that for every $p^0 \in N_\delta(p^*)$, $\lim_{t\to\infty} f_j(t; p^0) = p_j^*$, $j = 1, \ldots, n$.
3. p is said to possess *linear approximation stability* if the system

$$\dot{p}_j = g'(E_j) \sum_{r=1}^{n} \frac{\partial E_j}{\partial p_r}(p_r - p_r^*), j = 1, \ldots, n$$

   generates solution paths $f_j(p^0; t)$ such that for every p^0, $\lim_{t\to\infty} f_j(t; p^0) = p_j^*$, $j = 1, \ldots, n$ (derivatives are all evaluated at p^*).

Briefly, the relations among these concepts are as follows. Global stability implies local stability, and linear approximation stability implies local stability. Local stability implies neither global nor linear approximation stability, nor does global stability imply linear approximation stability. This chapter is primarily concerned with linear approximation stability.

Linear approximation stability under *tâtonnement* involves solving the Taylor series expansion,

$$\dot{p} = DAz,$$

where $\dot{p} = [\dot{p}_j]$, $j = 1, \ldots, n$, D is a diagonal matrix with $d_{ii} > 0$, $z = [p_j - p_j^*]$,

$$A = \left[\frac{\partial E_i}{\partial p_j} \right], i, j = 1, \ldots, n,$$

and

$$A^* = \left[\frac{\partial E_i}{\partial p_j} \right], i, j = 0, \ldots, n.$$

Recall that Walras's law and homogeneity impose the restriction that row and column sums in A^* are zero.

8.2 STABILITY UNDER WALRAS'S LAW AND HOMOGENEITY

As in Chapter 7, let

$$S_{A^*} = \left\{ B^* \mid B^* \in Q_{A^*} \text{ and } \sum_{j=0}^{n} b_{ij} = 0, i = 0, \ldots, n, \sum_{i=0}^{n} b_{ij} = 0, j = 0, \ldots, n \right\}.$$

Then A is said to be *sign stable* under Walras's law (W) and homogeneity (H) if $C^* \in S_{A^*}$ implies that C is a stable matrix, where $C = [c_{ij}]$, $i, j = 1, \ldots, n$. A is said to be sign D-stable under (W) and (H) if $C^* \in S_{A^*}$ implies that DC is a stable matrix, for every diagonal matrix D with positive diagonal entries. We first summarize the classical results concerning the (weak) gross substitute case, using McKenzie's (1960a) approach.

THEOREM 8.1. *Assume A^* is an irreducible weak gross substitute matrix, i.e., a matrix such that all goods are weak substitutes for one another, and the law of demand holds for all goods; and assume A^* satisfies (W),*

$$\sum_{i=0}^{n} a_{ij} = 0, j = 0, \ldots, n,$$

and (H),

$$\sum_{j=0}^{n} a_{ij} = 0, i = 0, \ldots, n.$$

Then $B^* \in S_{A^*}$ implies that B is a D-stable matrix, i.e., that A is sign D-stable under (W) and (H). Further, $B^* \in S_{A^*}$ implies that B^{-1} has all entries negative.

PROOF. Under the assumptions of the theorem, $B^* \in S_{A^*}$ implies $\sum_{j=1}^{n} b_{ij} \leq 0$, $i = 1, \ldots, n$, with strict inequality for some i, since $b_{00} < 0$ implies $b_{i0} > 0$ for some i. Thus $|b_{ii}| \geq \sum_{j \neq i} |b_{ij}|$ for $i = 1, \ldots, n$, with strict inequality for some i. Thus B is dominant (negative) diagonal and hence stable. Since dominant negative diagonal matrices preserve the dominant diagonal property when premultiplied by a (positive) diagonal matrix, B is D-stable as well. Finally, by the Perron-Frobenius theorem, $B^{-1} < 0$ if B is an irreducible stable weak gross substitute matrix.

Note that the proof utilized only the restriction on column sums in A^* (row sums could have been used instead); hence A is sign stable under (W), i.e., when only (W) but not (H) holds, as well as sign stable under (W) and (H). Generally speaking, sign stability under (W) and (H) does not imply sign stability under (W), but the weak gross substitute assumptions do guarantee this result. On the other hand, sign stability under (W) implies sign stability under (W) and (H).

Actually, a considerably stronger result can be obtained for the weak gross substitute case, namely, global stability of the system $\dot{p}_i = E_i(p)$, $i = 0,\ldots,n$.[1] We use McKenzie's proof (1960a).

THEOREM 8.2. *Assume $E_i(p)$ satisfies the weak gross substitute assumptions for all $p \geq 0$, i.e., $\partial E_i/\partial p_i < 0$, $i = 0,\ldots,n$, $\partial E_i/\partial p_j \geq 0$, $i \neq j$, $i, j = 0,\ldots,n$, evaluated at any $p \geq 0$, with $A^*(p) = [\partial E_j/\partial p_i]$, $i, j = 0,\ldots,n$, irreducible for any p. Then, if $E_i(t; p^0)$ is the solution path for E_i from the system $\dot{p}_i = E_i(p)$, $i = 0,\ldots,n$, $\lim_{t\to\infty} E_i(p^0; t) = 0$ for any $p^0 > 0$.*

PROOF. By Walras's law (W), $\Sigma_{i=0}^n p_i E_i(p) = 0$ for any $p \geq 0$; hence $\Sigma_{i=0}^n p_i \partial E_i/\partial p_j = -E_j(p)$ $j = 0,\ldots,n$. By homogeneity (H), $\Sigma_{j=0}^n p_j \partial E_i/\partial p_j = 0$. Let $a_{ij} = \partial E_i/\partial p_j$, so that (W) implies $\Sigma_{i=0}^n p_i a_{ij} = -E_j$ and (H) implies $\Sigma_{j=0}^n p_j a_{ij} = 0$. Let $P = \{i \mid E_i(p) > 0\}$, and let $V = 1/2\Sigma_{i\in P}(E_i(p))^2$. Then, under the assumption $\dot{p}_i = E_i(p)$ $i = 0,\ldots,n$,

$$\frac{dV}{dt} = \sum_{i\in P} E_i \sum_{j=0}^{n} a_{ij}E_j = \sum_{i\in P}\sum_{j\in P} E_i a_{ij} E_j + \sum_{i\in P}\sum_{j\notin P} E_i a_{ij} E_j,$$

where, since $a_{ij} \geq 0$ $i \neq j$, $E_j \leq 0$ $j \notin P$, $E_i > 0$ $i \in P$, the last sum is nonpositive.

Consider $\Sigma_{i\in P}\Sigma_{j\in P} E_i a_{ij} E_j$. Under weak gross substitutability, $\Sigma_{j\in P} a_{ij} p_j \leq 0$, $i \in P$ from (H) (with strict inequality for some $i \in P$ by irreducibility). From (W), $\Sigma_{i\in P} p_i a_{ij} \leq 0$ $j \in P$ (with strict inequality for some $j \in P$ by indecomposability). Thus $[a_{ij} + a_{ji}]$, $i \in P$ $j \in P$ has quasi-dominant diagonal. (In particular, $2p_i|a_{ii}| \geq \Sigma_{j\neq i\, j\in P} p_j(a_{ij} + a_{ji})$, $i \in P$). Hence $[a_{ij} + a_{ji}]$ has negative eigenvalues; i.e., $[a_{ij} + a_{ji}]$ is negative definite so that $\Sigma_{i\in P}\Sigma_{j\in P} E_i a_{ij} E_j < 0$, also. Hence $dV/dt < 0$ for $E_P \neq 0$, where E_P is the vector of excess demands, E_i, $i \in P$. But Walras's law $\Sigma_{i=0}^n p_i E_i(p) = 0$ together with $p_i > 0$ for $i = 0,\ldots,n$ imply that $E_p = 0$ if and only if $E_i(p) = 0$ for every $i = 0,\ldots,n$. Thus if

$E_i(p) < 0$ for some i, this implies $E_j(p) > 0$ for some j. Hence, by Liapunov's theorem, $\lim_{t \to \infty} E^i(p^0, t) = 0$, $i = 0, \ldots, n$.

Turning to linear approximation stability, when the assumptions of the gross substitute case are relaxed, the analysis becomes considerably more complicated. The issue that arises is that of determining whether complementarity is consistent with sign stability under (W) and (H). For the special case of "sign symmetry," where $\operatorname{sgn} a_{ij} = \operatorname{sgn} a_{ji}$ $i \neq j$, it can be shown that complementarity cannot occur at a sign stable equilibrium (see Quirk 1970).

Lemma 8.1. *Assume A^* is sign symmetric* ($\operatorname{sgn} a_{ij} = \operatorname{sgn} a_{ji}$ $i \neq j$). *Then if A is sign stable under (W) and (H), $B^* \in S_{A^*}$ implies that B is quasi-negative definite, i.e., that $x'Bx < 0$ for $x \neq 0$.*

Proof. A sign stable under (W) and (H) means $B^* \in S_{A^*}$ implies B is a stable matrix. Consider $C^* = B^* + B^{*\prime} = [b_{ij} + b_{ji}]$. Then, under sign symmetry, C^* satisfies $\operatorname{sgn} c_{ij} = \operatorname{sgn}(b_{ij} + b_{ji}) = \operatorname{sgn} a_{ij}$; hence $C^* \in Q_A^*$. Further, $B^* \in S_{A^*}$ implies $\Sigma_{i=0}^n (b_{ij} + b_{ji}) = 0$, $j = 0, \ldots, n$ and $\Sigma_{j=0}^n (b_{ij} + b_{ji}) = 0$ $i = 0, \ldots, n$; hence $C^* \in S_{A^*}$. Thus if A is sign stable under (W) and (H), $B^* \in S_{A^*}$ implies $B + B'$ is a stable matrix. But $B + B'$ is symmetric, so $B + B'$ is negative definite, which in turn implies that $x'Bx < 0$ for $x \neq 0$; i.e., B is quasi-negative definite.

Lemma 8.2. *Let A^* satisfy $\Sigma_{i=0}^n a_{ij} = 0$ $j = 0, \ldots, n$, $\Sigma_{j=0}^n a_{ij} = 0$ $i = 0, \ldots, n$. Then A is quasi-negative definite if and only if every $n \times n$ principal submatrix of A^* is quasi-negative definite.*

Proof. Let $Q(h^*) = h^{*\prime} A^* h' = \Sigma_{i=0}^n \Sigma_{j=0}^n a_{ij} h_i h_j$, where $h^* = (h_0, h_1, \ldots, h_n)$.

Assume that A is quasi-negative definite, i.e., that $h'Ah = \Sigma_{i=1}^n \Sigma_{j=1}^n a_{ij} h_i h_j < 0$, for $h \neq 0$, where $h = (h_1, \ldots, h_n)$. We wish to show that every $n \times n$ principal submatrix of A^* is quasi-negative definite given that $\Sigma_{i=0}^n a_{ij} = 0$ $j = 0, \ldots, n$, $\Sigma_{j=0}^n a_{ij} = 0$ $i = 0, \ldots, n$.

We first show that $Q(h^*) \leq 0$ for every h^*. Clearly if $h_0 = 0$, this follows directly from the quasi-negative definiteness of A. Hence assume $h_0 \neq 0$, and write $Q(h^*)$ as

$$Q(h^*) = a_{00} h_0^2 + h_0 \sum_{i=1}^n a_{i0} h_i + h_0 \sum_{j=1}^n a_{0j} h_j + \sum_{i=1}^n \sum_{j=1}^n a_{ij} h_i h_j,$$

where $a_{i0} = -\Sigma_{j=1}^n a_{ij}$, $a_{0j} = -\Sigma_{i=1}^n a_{ij}$, $a_{00} = \Sigma_{i=1}^n \Sigma_{j=1}^n a_{ij}$. Thus, $Q(h^*) = \Sigma_{i=1}^n \Sigma_{j=1}^n a_{ij}(h_0^2 - h_i h_0 - h_0 h_j + h_i h_j)$. Let $y_i = h_i / h_0$ so that

$1/h_0^2 Q(h^*) = \Sigma_{i=1}^n \Sigma_{j=1}^n a_{ij}(1 - y_i - y_j + y_i y_j)$, i.e., $1/h_0^2 Q(h^*) = \Sigma_{i=1}^n \Sigma_{j=1}^n a_{ij}(1 - y_i)(1 - y_j)$. But now by quasi-negative definiteness of A, $1/h_0^2 Q(h^*) \leq 0$ and $Q(h^*) < 0$ except when $y_i = 1$ $i = 1, \ldots, n$, i.e., when $h_i = h_0$ $i = 1, \ldots, n$.

Next, note that the quadratic form associated with any $n \times n$ principal submatrix of A^* is obtained from $Q(h^*)$ by setting $h_k = 0$ for the index k deleted from $\{0, 1, \ldots, n\}$ in the index set associated with the submatrix. Since $Q(h^*) < 0$ except when $h_i = h_0$ $i = 1, \ldots, n$ and $h_k = 0$, it follows that the quadratic form associated with the principal submatrix is negative except when $h^* = 0$, which establishes the lemma.

LEMMA 8.3. *Assume A is quasi-negative definite. Then every ith-order principal minor of A has sign $(-1)^i$ $i = 1, \ldots, n$.*

PROOF. Quasi-negative definiteness implies that A is stable, since $A + A'$ is negative definite; thus the Liapunov criterion applies (i.e., A is stable if and only if there exists a symmetric positive definite matrix B such that $BA + A'B$ is negative definite—here the matrix B is simply the identity matrix). In fact, every principal submatrix of A is stable by the same argument. Since stability implies that the Routh-Hurwitz conditions are satisfied, the determinant of any ith-order principal minor has sign $(-1)^i$, $i = 1, \ldots, n$.

COROLLARY. *Assume that A^* is sign symmetric and that A is stable under (W) and (H). Then $B^* \in S_{A^*}$ implies that every ith-order principal minor of B^* has sign $(-1)^i$ $i = 1, \ldots, n$.*

From the above, the result concerning sign symmetry and complementarity follows directly.

THEOREM 8.3. *Assume that A^* is sign symmetric and irreducible. Then A is sign stable under (W) and (H) and sign D-stable under (W) and (H) if and only if $a_{ii} < 0$ $i = 0, \ldots, n$ and $a_{ij} \geq 0$ $i \neq j$ $i, j = 0, \ldots, n$.*

PROOF. Sufficiency follows from Theorem 8.1. To show necessity, we note from the corollary to Lemma 8.3 that all first-order principal minors of A^* must be negative; hence $a_{ii} < 0$ $i = 0, \ldots, n$. Assume that for some $i \neq j$, $a_{ij} < 0$, $a_{ji} < 0$. Since A is sign stable under (W) and (H), $B^* \in S_{A^*}$ implies $B^* + B^{*\prime} \in S_{A^*}$; hence $B + B'$ is stable. With no loss of generality, assume that $C^* \in S_{A^*}$ is a symmetric matrix, with C stable.

Let $\alpha_i = c_{ii} + c_{ij} = c_{ii} + c_{ji}$, and let $\alpha_j = c_{jj} + c_{ji} = c_{jj} + c_{ij}$, where $\alpha_i < 0$, $\alpha_j < 0$. Choose M^* with $m_{ii} = \alpha_i^2/\alpha_i + \alpha_j + \varepsilon$, $m_{jj} = \alpha_j^2/\alpha_i + \alpha_j + \varepsilon$, $m_{ij} = \alpha_i \alpha_j/\alpha_i + \alpha_j - \varepsilon$, with all other entries in M^* identical to those in C^*, where ε is an arbitrarily small positive number.

Then $m_{ii} + m_{ij} = m_{ii} + m_{ji} = \alpha_i$, $m_{jj} + m_{ij} = m_{jj} + m_{ji} = \alpha_j$, and $M^* \in S_{A^*}$. However, $m_{ii}m_{jj} - m_{ij}m_{ji} = \varepsilon(\alpha_i + \alpha_j) < 0$, which violates the condition that all second-order principal minors of M^* must be positive.

From theorem 8.3 it follows that for the case in which (W) and (H) hold and substitutability, complementarity, and independence are symmetric relations, stability can be proved from the sign pattern of A^* alone in only the (weak) gross substitute case. Of course, if sufficiently strong quantitative restrictions are imposed on the competitive model, stability might still be proved under sign symmetry in the presence of complementarity. An extreme example of this is in the case of a pure trade economy in which no trade occurs at equilibrium, so that the coefficient matrix A^* is symmetric and negative semi-definite of rank n. See Arrow and Hurwicz (1959) and Quirk and Saposnik (1968) for a discussion of this case.

As it turns out, there are classes of cases beyond the gross substitute case in which stability of the competitive equilibrium can be established by using a qualitative approach. As in Chapter 7, the analysis utilizes the concept of a free entry in A under (W) and (H), defined as an entry a_{rs} in A such that given any real number $M > 0$, there exists $B^* \in S_{A^*}$ such that $|b_{rs}| > M|b_{ij}|$ for every $i \neq r$ and $j \neq si$, $j = 1, \ldots, n$. Then a_{rs} is a free entry in A under (W) and (H) if and only if $a_{rs}a_{r0} < 0$ and $a_{rs}a_{0s} < 0$ with $a_{rs}a_{00} > 0$.

We begin our analysis with the case in which A (but not A^*) is gross substitute so that $a_{ii} < 0$ $i = 1, \ldots, n$, with $a_{ij} > 0$ $i \neq j$ $i, j = 1, \ldots, n$. It is also assumed that $a_{00} < 0$. From the earlier analysis, it is known that if $a_{i0} > 0$ $i = 1, \ldots, n$ or if $a_{0j} > 0$ $j = 1, \ldots, n$, then A is sign stable under (W) and (H). Theorem 8.4 below identifies one other case of (W) and (H) sign stability.

THEOREM 8.4. *Assume A is gross substitute and that $a_{00} < 0$; in addition, assume $a_{0i}a_{i0} < 0$ for all i and there is only one index k for which a_{0k} is negative. Then A is sign stable under (W) and (H).*

PROOF. We will show that under the conditions of the theorem, A is dominant diagonal, i.e., that there exist positive constants $d_1, \ldots, d_n$ such that

$$d_i|a_{ii}| > \sum_{j \neq i} d_j|a_{ij}| j = 1, \ldots, n.$$

We first show that $\operatorname{sgn} \det A = (-1)^n$. Without loss of generality, let $a_{10} < 0$, $a_{i0} > 0$ $i = 2, \ldots, n$, with $a_{01} > 0$, $a_{0j} < 0$ $j = 2, \ldots, n$, so that A^*

has the sign pattern

$$\operatorname{sgn} A^* = \begin{bmatrix} - & + & - & - & - \\ - & - & + & + & + \\ + & + & - & + & + \\ + & + & + & - & + \\ + & + & + & + & - \end{bmatrix},$$

with rows and columns numbered $0, 1, \ldots, n$. Consider the principal minor of A^*, A_{11}^*, formed by deleting row 1 and column 1. Then A_{11}^* has the sign pattern

$$\operatorname{sgn} A_{11}^* = \begin{bmatrix} - & - & - & - \\ + & - & + & + \\ + & + & - & + \\ + & + & + & - \end{bmatrix},$$

with all row sums positive by (W) and the sign pattern of A^*. But this implies $\operatorname{sgn}\det A_{11}^* = (-1)^n$ by Lemma 7.6. Since all $n \times n$ cofactors in A^* are equal by theorem 7.10, then $\operatorname{sgn}\det A = (-1)^n$, as well.

To obtain positive constants $d_1, \ldots, d_n$ that prove the quasi-dominant diagonal property of A, choose $d_1 = 1$ and choose $d_2, \ldots, d_n$ by solving the system

$$\sum_{j=2}^{n} a_{ij} d_j = -a_{i1} \; i = 2, \ldots, n.$$

Note that because $a_{i0} > 0$, $a_{i1} > 0$, $i = 2, \ldots, n$, the matrix $[a_{ij}]$, $i, j = 2, \ldots, n$ is dominant diagonal, and hence this matrix has determinant of sign $(-1)^{n-1}$ with all entries in its inverse negative by the Frobenius theorem. Let Δ_{rs} denote the cofactor of a_{rs} in A, and let Δ_{rs}^{ij} denote the cofactor of a_{ij} in Δ_{rs}. Then, solving the above system, we obtain

$$d_j = -\sum_{i=2}^{n} \frac{\Delta_{11}^{ij} a_{i1}}{\Delta_{11}} j = 2, \ldots, n.$$

Thus $d_j > 0$ $j = 2, \ldots, n$, since $\Delta_{11}^{ij}/\Delta_{11} < 0$ $i, j = 2, \ldots, n$, while $a_{i1} > 0$ $i = 2, \ldots, n$.

Under the choice of $d_1 = 1$, we have $\Sigma_{j=1}^{n} d_j a_{ij} = 0 \; i = 2,\ldots,n$, and under the sign pattern of A^*, this implies

$$d_i|a_{ii}| = \sum_{\substack{j=1\\ j\neq 1}}^{n} d_j|a_{ij}| i = 2,\ldots,n.$$

We will show that $d_1|a_{11}| > \Sigma_{j=2}^{n} d_j|a_{1j}|$. The proof utilizes the identity

$$\Delta_{i1} = -\sum_{j=2}^{n} a_{1j}\Delta_{11}^{ij} i = 2,\ldots,n.$$

From the above,

$$\sum_{j=2}^{n} d_j a_{1j} = -\sum_{j=2}^{n}\sum_{i=2}^{n} a_{1j}\frac{\Delta_{11}^{ij}a_{i1}}{\Delta_{11}} = -\sum_{i=2}^{n}\frac{a_{i1}}{\Delta_{11}}\sum_{j=2}^{n} a_{1j}\Delta_{11}^{ij} = \sum_{i=2}^{n}\frac{a_{i1}\Delta_{i1}}{\Delta_{11}}$$

But $\det A = \Sigma_{i=1}^{n} a_{i1}\Delta_{i1}$, with sgn $\det A = (-1)^n$ and sgn $\Delta_{11} = (-1)^{n-1}$; hence,

$$0 > \frac{\det A}{\Delta_{11}} = a_{11} + \sum_{i=2}^{n} a_{i1}\frac{\Delta_{i1}}{\Delta_{11}} = > |a_{11}| > \sum_{i=2}^{n} a_{i1}\frac{\Delta_{i1}}{\Delta_{11}}$$

$$= > d_1|a_{1j}| > \sum_{j=2}^{n} d_j|a_{ij}|.$$

Choosing $d_1 = 1 - \varepsilon$, where $\varepsilon > 0$ is arbitrarily small, we obtain

$$d_i|a_{ii}| > \sum_{j\neq 1} d_j|a_{ij}| i = 1,\ldots,n,$$

and A is stable. Since the proof utilizes only qualitative information together with (W) and (H), this implies that A is sign stable under (W) and (H).

It might be noted that the case taken up in Theorem 8.4 is a case in which sign stability under (W) does not hold—the quantitative restrictions imposed by (W) and by (H) both have to be utilized in the proof. This case would perhaps not be of so much interest in itself except that when $a_{ij} \neq 0 \; i,j = 0,\ldots,n$, and when A is gross substitute, it turns out that this is the only case (other than the "mirror image" case) in which sign stability under (W) and (H) holds and sign stability under (W) or sign stability under (H) does not hold. That is, except for the case in which the *numeraire* row (or column) is gross substitute and the case of Theorem 8.3, there are no other cases in which sign stability under (W) and (H) can be proved when A is gross substitute and $a_{ij} \neq 0$ $i, j = 0, \ldots, n$. The proof of this uses the notion of a free entry under (W) and (H) as defined earlier.

LEMMA 8.4. *Assume A^* satisfies $a_{ij} \neq 0$, $i, j = 0, \ldots, n$, with $a_{ii} < 0$, $i = 0, \ldots, n$ and $a_{ij} > 0$ $i \neq j$, $i, j = 1, \ldots, n$, $(n > 3)$. Then if A is sign stable under (W) and (H), $a_{0k} > 0$, $a_{k0} > 0$, for some $k \neq 0$ implies that if $a_{0t} < 0$ for some $t \neq 0$, then $a_{i0} > 0$ $i = 1, \ldots, n$.*

PROOF. If the matrix A is stable under (W) and (H), then A has signed determinant (of sign $(-1)^n$) under (W) and (H). Consider the cofactor A^*_{kk} in A^* formed by deleting the kth row and kth column of A^*. Note that $a_{kj} > 0$ $j \neq k$ and $a_{ik} > 0$ $i \neq k$, $i, j = 1, \ldots, n$, with $a_{kk} < 0$. By Lemma 7.3, A^*_{kk} has negative off-diagonal entries in at most one row or column, so that $a_{0t} < 0$ for some $t \neq 0$ implies $a_{i0} > 0$ $i = 1, \ldots, n$.

LEMMA 8.5. *Under the conditions of Lemma 8.4, if $a_{0k}a_{k0} < 0$ for $k = 1, \ldots, n$, with $n > 3$, and A is sign stable under (W) and (H), then either $a_{i0} < 0$ for at most one index $i \neq 0$ or $a_{0j} < 0$ for at most one index $j \neq 0$.*

PROOF. Without loss of generality, assume $a_{10} < 0$, $a_{20} < 0$ and $a_{03} < 0$, $a_{04} < 0$ (implies $a_{01} > 0$, $a_{02} > 0$, $a_{03} > 0$, $a_{04} > 0$). Consider the cofactor A^*_{11} formed by deleting row and column 1 from A^*. Within this cofactor, the entries $a_{22}, \ldots, a_{nn}$ are free entries, as are a_{03}, a_{04}. Choose $C^* \in Q_A$, with $c_{22} = \cdots c_{nn} = -3$, $c_{00} = -\delta$, $c_{20} = -1$, $c_{42} = 1$, $c_{04} = -1$, $c_{30} = 2$, all other entries in C having an absolute value of δ, where δ is an arbitrarily small positive number. Then row sums in C^*_{11} are all negative, the sum of entries in column 0 of C^*_{11} is positive, and all other column sums in C^*_{11} are negative; hence $C^* \in S_{A^*}$. Now C^*_{11} is dominated by the term $(c_{04}c_{42}c_{20}c_{55} \ldots c_{nn})$, which has sign $(-1)^{n+1}$, which implies $\operatorname{sgn}\det C = (-1)^{n+1}$; hence C is not stable, and thus A is not sign stable under (W) and (H).

LEMMA 8.6. *Under the conditions of Lemma 8.4, if A is sign stable under (W) and (H), then $a_{i0} < 0$ for any $i \neq 0$ implies $a_{0i} > 0$, with $n > 3$.*

PROOF. If there exists a k such that $a_{0k} > 0$, $a_{k0} > 0$, the proof follows immediately from Lemma 8.5. Hence assume $a_{0k} > 0$ implies $a_{k0} < 0$, $k = 1, \ldots, n$. Because $a_{00} < 0$, there exist r, s such that $a_{r0} > 0$, $a_{0s} > 0$. Assume that $a_{10} < 0$, $a_{01} < 0$. Let A^*_{rr} denote the cofactor formed by deleting row and column r from A^*. Then in A^*_{rr}, all negative entries except those appearing in row 0 are free entries. Hence, the terms $a_{00}a_{11}a_{22}\ a_{r-1,r-1}a_{r+1,r+1}\ a_{nn}$ and $-a_{01}a_{10}a_{22}\ a_{r-1,r-1}a_{r+1,r+1}\ a_{nn}$ both appear in the expansion of A^*_{rr}, and both contain $n-1$ free entries and are of opposite sign. Hence A does not have signed determinant under (W) and (H), and A is not sign stable under (W) and (H).

COROLLARY. *Under the conditions of Lemma* 8.4, *if A is sign stable under* (W) *and* (H), *with* $n > 3$, *then either*

(i) $a_{i0} > 0\ i = 1,\dots,n$; *or*
(ii) $a_{0j} > 0\ j = 1,\dots,n$; *or*
(iii) $a_{i0}a_{0i} < 0\ i = 1,\dots,n$, *and* $a_{i0} < 0$ *for at most one index i or* $a_{0j} < 0$ *for at most one index j.*

PROOF. From Lemma 8.4, if $a_{0k} > 0$, then $a_{k0} > 0$ implies (i) or (ii); hence either $a_{0k}a_{k0} < 0$ for all k, which, by Lemma 8.4, implies (iii), or $a_{0k} < 0$, $a_{k0} < 0$ for some k, which is excluded by Lemma 8.6.

THEOREM 8.5. *Assume* A^* *satisfies* $a_{ij} \neq 0\ i,j = 0,\dots,n$, *with* $n > 3$, $a_{ii} < 0\ i = 0,\dots,n$, $a_{ij} > 0\ i \neq j\ i,j = 1,\dots,n$. *Then A is sign stable under* (W) *and* (H) *and sign D-stable under* (W) *and* (H) *if and only if either*

(i) $a_{i0} > 0\ i = 1,\dots,n$; *or*
(ii) $a_{0j} > 0\ j = 1,\dots,n$; *or*
(iii) $a_{i0}a_{0i} < 0\ i = 1,\dots,n$, *and* $a_{i0} < 0$ *for at most one index i or* $a_{0j} < 0$ *for at most one index j.*

In the earlier work by Hicks and Morishima, attention was centered on the special case where the *numeraire* is a substitute for all other goods and conversely and the law of demand holds for all goods; in this case, the matrix A^* has negative diagonal entries, and the zero row and column of A^* have positive off-diagonal entries. For this class of competitive environments, necessary and sufficient conditions for establishing sign stability under (W) and (H) can be derived as follows. Under the assumption that A^* has a negative diagonal, with off-diagonal entries in row and column zero being positive, then the free entries in A under (W) and (H) are the negative entries in A, including the diagonal entries. By the Routh-Hurwitz conditions, a necessary condition for stability of A is that $\operatorname{sgn}\det A = (-1)^n$. A necessary condition for A to be sign stable under (W) and (H) is that A have signed determinant under (W) and (H) (of sign $(-1)^n$). Thus in the case where the *numeraire* is a substitute for all other goods and conversely and the law of demand holds for all goods, the results of Chapter 7 apply directly to the stability problem.

In particular, we have seen that by Theorem 7.11, a necessary and sufficient condition for sign solvability under (W) and (H) is that A be sign Hicksian under (W) and (H). In turn, this occurs if and only if (i) $a_{ij} < 0$ $i \neq j$ implies that a_{ij} appears only in nonpositive cycles in A; and (ii) condition Q′ holds.

LEMMA 8.7. *If the numeraire is a substitute for all other goods and conversely and the law of demand holds for all goods, then A is sign stable under (W) and (H) only if A is sign Hicksian under (W) and (H).*

Given that the *numeraire* is a substitute for all other goods and conversely and that the law of demand holds for all goods, the links between dynamic stability and Hicksian stability are clearly very close in a qualitatively specified environment. In addition, from Lemma 8.7, it follows that sign stability under (W) and (H) cannot be proved if A contains any pair of pure complementary goods i and j, i.e., goods such that i is a complement for j and j is a complement for i.

In the pure sign stable case, discussed in Chapter 2, cycles of length greater than two must be excluded, because of potential violations of the Routh-Hurwitz equivalent conditions for stability, i.e., $k_i > 0$ $i = 1,\ldots,n$, where $k_i = (-1)^i$ times the sum of all ith-order principal minors of the matrix, and the determinantal conditions

$$\begin{vmatrix} k_1 & k_3 \\ 1 & k_2 \end{vmatrix} > 0, \begin{vmatrix} k_1 & k_3 & k_5 \\ 1 & k_2 & k_4 \\ 0 & k_1 & k_3 \end{vmatrix} > 0, \ldots.$$

From the Maybee determinant formula (A2.2), positive cycles of length i enter into k_i, $i = 2,\ldots,n$ with "incorrect" sign $((-1)^{i=1})$; hence positive cycles of all lengths must be excluded in the pure sign stable case. Negative cycles enter the k_i conditions with proper sign, but negative cycles of length three or greater enter one or more Routh-Hurwitz determinantal conditions with "incorrect" sign, and hence also must be excluded under sign stability. The point is that if a matrix is sign stable, then every noncanceled term in both the k_i conditions and the determinantal conditions must have the "correct" sign—a single term of incorrect sign invalidates pure sign stability.

When (W) and (H) are imposed on A^*, this introduces quantitative restrictions on the values of the elements of A, and hence on the values of the cycles that appear in A. In particular, when the *numeraire* is a substitute for all other goods and conversely and the law of demand applies to all goods, (W) and (H) impose restrictions on the values that can be taken on by the positive entries in A, but not on the values that can be taken on by the negative elements in A. Thus, since negative off-diagonal entries in A can appear only in negative cycles if sign stability under (W) and (H) is to be established, we can immediately arrive at a further necessary condition for such sign stability, namely, that A cannot contain any negative cycles of length greater than two.

These necessary conditions for (W) and (H) sign stability turn out to be sufficient as well. Theorem 8.16 provides a summary.

THEOREM 8.16. *If the numeraire is a substitute for all other goods and conversely and the law of demand applies for all goods, necessary and sufficient conditions for A to be sign stable under* (W) *and* (H) *are*

(1) *A is sign Hicksian under* (W) *and* (H), *i.e.,*
  (i) $a_{ij} < 0$ $i \neq j$ *implies that* a_{ij} *appears only in nonpositive cycles in* A, *and*
  (ii) *condition* Q' *holds; and*

(2) *A contains no negative cycles of length greater than two.*

PROOF. Necessity of (1) follows from the fact that if A is sign stable under (W) and (H), then, by the Routh-Hurwitz conditions, A has signed determinant under (W) and (H), with sign $(-1)^n$, which, by Lemma 8.7, implies that A is sign Hicksian under (W) and (H). Necessity of (2) follows from the fact that (W) and (H) impose no quantitative constraints on the values of negative off-diagonal entries in A, and hence no constraints on the values of the negative cycles in A. Since, by the pure sign stability theorem, negative cycles of length greater than two lead to the violation of one or more of the Routh-Hurwitz determinantal conditions, this also applies in the case where (W) and (H) hold.

For sufficiency, first note that if A contains only nonnegative cycles and also satisfies (1), then A is sign stable under (W) and (H), since a necessary and sufficient condition for $B \in Q_A$ to be stable in this case is that B be Hicksian, i.e., that (1) hold. If A contains only nonpositive cycles, then (2) is the condition for A to be a pure sign stable matrix, so A is sign stable under (W) and (H).

Suppose that A contains some negative cycles of length two, together with positive cycles of various lengths. Looking just at the negative cycles, first note that under (1), since A is sign Hicksian under (W) and (H), negative cycles of all orders enter principal minors with correct sign, so that negative cycles cause no problems so far as the Routh-Hurwitz $k_i > 0$ conditions are concerned. The Routh-Hurwitz determinantal conditions are more interesting. As demonstrated in the discussion of pure sign stable matrices in Chapter 2, negative cycles of length two are acceptable, while negative cycles of length greater than two are not. Since there are terms in the Routh-Hurwitz determinantal conditions that involve negative cycles of length two with incorrect sign, it must be that all such terms are canceled out by corresponding terms appearing elsewhere in the expansion of the determinant, and this applies for all of the determinantal conditions. For example, for a 3×3 matrix, the Routh-Hurwitz conditions $k_1 k_2 - k_3 > 0$ can be written as $-(a_{11} + a_{22} + a_{33})(\Delta_{12} + \Delta_{13} + \Delta_{23}) + [a_{11}\Delta_{23} + a_{22}\Delta_{13} + a_{33}\Delta_{12} + a_{12}a_{23}a_{31} + a_{13}a_{32}a_{21} - 2a_{11}a_{22}a_{33}) > 0$, where $\Delta_{12} = a_{11}a_{22} - a_{12}a_{21}$, etc.

After cancellations of like terms, and given that the matrix is Hicksian, all the terms involving cycles of length two, the Δ_{ij} terms, have correct sign (positive). Any positive cycle of length three has correct sign, and negative three cycles have the incorrect sign for the determinantal condition. The fact that the sign stability theorem permits the appearance of negative cycles of length two implies that in any determinantal condition, after cancellations, such cycles always appear in second-order principal minors, which multiply other principal minors, with the correct (positive) sign, so long as the matrix is Hicksian. This applies whether or not the matrix is a pure sign stable matrix. Thus, the existence of negative cycles of length two in a Hicksian matrix that also contains positive cycles creates no problems with the Routh-Hurwitz determinantal conditions, or with the $k_i > 0$ conditions.

In the case of principal minors containing only positive cycles, the fact that condition (1) holds means that all of the $k_i > 0$ conditions are satisfied. It can also be shown that, after cancellations of like terms with opposite signs, every remaining term in a Routh-Hurwitz determinantal condition is positive, given that all cycles are nonnegative and the matrix is Hicksian.[2] Terms in the Routh-Hurwitz determinantal conditions involving cycles of length two are positive by the earlier argument. Under conditions (1) and (2), all of the remaining terms in any Routh-Hurwitz determinantal conditions involve only nonnegative cycles. Hence under (1) and (2), all Routh-Hurwitz determinantal conditions and $k_i > 0$ conditions are satisfied, so the theorem holds.

Finally, as in the discussion of comparative statics results, one case of interest is the case where A is a full matrix. Here a particularly strong and simple result holds.

THEOREM 8.7. *If the numeraire is a substitute for other goods and conversely, the law of demand holds for all goods, and A is a full matrix*

$(a_{ij} \neq 0\, i\, j = 1, \ldots, n)$, with $n > 2$, then A is sign stable under (W) and (H) if and only if A^* is a Metzler (gross substitute) matrix.

PROOF. If there exists $a_{ij} < 0$ in A, $i \neq j$, then for $n > 2$, a_{ij} will appear in a nonzero cycle, $a_{ij} a_{kj} a_{ki}$. If the cycle is positive, A violates condition (1(i)) of Theorem 8.6; if the cycle is negative, A violates condition (2) of Theorem 8.16. On the other hand, if all off-diagonal elements in A^* are positive, $B^* \in S_{A^*}$ implies that B is Metzler and dominant diagonal, and hence stable.

For $n = 2$, of course, pure sign stable matrices, such as $\begin{bmatrix} - & + \\ - & - \end{bmatrix}$ or $\begin{bmatrix} - & - \\ + & - \end{bmatrix}$, are full matrices that are sign stable under (W) and (H), which accounts for the qualification in the theorem.

To summarize, Theorems 8.6 and 8.7 identify the classes of matrices that can be proven stable on the basis of their patterns of complementarity-substitutability properties, along with (W) and (H), assuming the *numeraire* is a substitute for all goods and conversely and the law of demand holds for all goods. Pairs of pure complementary goods are excluded; further, if j is a complement for i, then i is either a substitute for j or independent of j, and no indirect effects (nonzero chains of length greater than one) are permitted linking i to j.

The Hicks-Morishima approach of analyzing the comparative statics properties of the competitive equilibrium through the complementarity-substitutability relations among commodities gives rise to classes of cases beyond the gross substitute and pure sign stable cases, where restrictive comparative statics theorems can be derived, and (dynamic) stability can be proved. While there are other possible approaches to the comparative statics of the competitive equilibrium, still the qualitative approach has the advantage that the results obtained are less sensitive to the problems posed by volatility of the parameters of the competitive model due to changes in technology, tastes, and institutions.

The analysis presented here helps to clarify the links between Hicksian stability and dynamic stability in the context of the competitive model; it also highlights the problems posed for the derivation of restrictive comparative statics results and proof of dynamic stability of a *tâtonnement* mechanism by the presence of pure complementary goods. As long as negative cycles of length greater than two are excluded, dynamic stability and Hicksian stability are functionally equivalent in a qualitative setting in which the *numeraire* is a substitute for all other goods and conversely and the law of demand holds for all goods, in the sense that each is necessary and sufficient for proving the existence of the other. The analysis indicates that there are severe limitations on the extent to which the gross substitute and stable cases can be extended, the most important of which are the limitations placed on the presence of complementary relations among goods. The restrictive class of qualitatively specified matrices for which stability can be proved has led to the investigation of systems in which additional quantitative information is incorporated into the analysis in order to establish stability. Examples of this approach include the work of Ohyama (1972) and Mukherji (1972).

Ohyama introduces the notion of a G_1-Metzlerian matrix A^* as one that satisfies

(i) $a_{0j} > 0 \; j = 1, \ldots, n$, $a_{i0} > 0 \; i = 1, \ldots, n$ (and $a_{00} < 0$); and

(ii) there exists an $n \times n$ stochastic positive definite matrix G such that GA is a Metzlerian matrix, where $g_{ij} \geqq 0 \; i, j = 1, \ldots, n$, with $\Sigma_{j=1}^{n} g_{ij} = 1$.

Similarly, A^* is a G_2-Metzlerian matrix if A^* satisfies condition (i) for a G_1-Metzlerian matrix and

(ii′) there exists an $n \times n$ stochastic positive definite matrix G such that $G(A + A')$ is a Metzlerian matrix, where, again, $g_{ij} \geqq 0$ $i, j = 1, \ldots, n$, with $\Sigma_{j=1}^{n} g_{ij} = 1$.

Then the Ohyama (1972) results are the following (proofs are omitted).

THEOREM 8.8. *If A^* is a G_1-Metzlerian matrix and row and column sums in A^* are zero, then A is a stable matrix.*

THEOREM 8.9. *If A^* is a G_2-Metzlerian matrix and row and column sums in A^* are zero, then A is a D-stable matrix.*

Thus the quantitative restrictions imposed by the assumptions of G_1- or G_2-Metzlerian matrices are sufficient to guarantee local stability of the competitive equilibrium whatever the substitutability-complementarity properties of A^*.

Mukherji's (1972) treatment deals with a generalized gross substitute matrix, defined as a matrix A^* such that there exists a nonsingular matrix S, $S \geq 0$ where $S[A^* + A^{*\prime}]S^{-1}$ is an irreducible Metzlerian matrix. The main result of the Mukherji paper is the following.

THEOREM 8.10. *If A^* is a generalized gross substitute matrix and if row and column sums in A^* are zero, then A is a stable matrix.*

Finally we should note that another example of a competitive system exhibiting sign stability under (W) and (H) is the case of a system in which A is sign stable in purely qualitative terms, e.g., where A has negative diagonal and nonpositive cycles of length two and all cycles of length greater than two are zero, assuming only that A^* has a sign pattern consistent with (W) and (H).

Notes

Chapter 1

1. The response of the government was to subdivide the steel industry classification into separate steel subindustries, reestimate, and then use the revised model for stockpile planning. The resulting stockpiling plan was in fact implemented for a number of strategic materials during the 1950s. Fortunately, the dubious accuracy of the projected stockpiling requirements was never put to the test by war, and the government ultimately sold most of its strategic stockpiles.

2. In the early 1960s, one of the authors of this book was involved in the early stages of a research project designed to fill a gap in the historical record of the U.S. economy. In the distant past, a fire at the Bureau of the Census had destroyed the records of an early Census of Manufactures, leading economic historians to seek out alternative data sources to estimate national income and gross national product for that period. The approach adapted in that research project was to model the early nineteenth-century U.S. economy as a dynamic input-output system, using estimates of coefficients from a variety of historical sources. The model's results were of mixed quality. The GNP estimate was quite close to the estimates that had been produced by economic historians using quite different methodologies. But, 90 percent of the calculated output of the U.S. economy from the model was steamboats! As it turned out, researchers found that relatively minor adjustments in their coefficient estimates produced a more sensible result. The sensitivity of those calculations to small errors further emphasizes the hazards present in drawing sweeping conclusions from data-based quantitative models.

3. An exception is the consideration of interval measures in Chapter 3.

4. A well-known sufficient condition for asymptotic convergence of y to y^* in (1.7) for arbitrary initial conditions is that all characteristic roots of DA have negative real parts. Such matrices are called stable.

5. In Berndt (1991, 553) the four behavioral coefficients of Klein's (macroeconomic) Model I are estimated by seven alternative methods as a pedagogical illustration. The smallest difference between the largest and smallest values found was 50% of the smaller value. In two cases, coefficients of the same order of magnitude were estimated with opposite signs. The alternative estimates were derived from the same data. A qualitative analysis of Klein's Model I is set forth in Chapter 4.

Chapter 2

1. The Jacobian matrix for the Oil Market Simulation Model presented in Chapter 4 is 19×19. The determinant of this matrix has over 10^{17} terms. Even so, for the applied model reviewed here, all of the entries in the 19×19 inverse Jacobian matrix could be signed.

2. The corresponding columns can be swapped as well if need be to keep the main diagonal terms nonzero.

3. In selecting a notation for sign patterns, there is a tension between using "+," "−," and "0," which are the natural symbols for the concepts of positive, negative, and zero, and using "1," "−1," "−1," and "0," which are the corresponding integer results of applying the "sgn" function in mathematics (i.e., $\operatorname{sgn} a = 1$ for $a > 0$, $\operatorname{sgn} a = -1$ for $a < 0$, and $\operatorname{sgn} a = 0$ for $a = 0$). Both notations are used in this book, depending upon the context of the discussion. If algorithmic or manipulative arguments involving signs are presented, then the results of using the sgn function $(1, -1, 0)$ will be used in displaying the results of the manipulations. When the discussion is descriptive calling for a schematic of the sign pattern of an array, the symbols "+," "−," and "0" will be used.

Chapter 3

1. Since each entry of an inverse matrix can be found as the ratio of two determinants, with the determinant of the matrix being inverted in the denominator of the fraction, the analysis results can be applied to the issue of finding each entry of A^{-1}.

2. In fact, for B in SF1 and b_{ij}^m, an entry of the mth Boolean power of B, there exists a path $p(j \to i)$ of length m or less (i.e., a path expressed with m-many or fewer arrows) if and only if $b_{ij}^m = 1$. In manipulating entries of B when taking powers, standard matrix operations are utilized with Boolean addition and multiplication, as $u + v = \max(u, v)$ and $uv = \min(u, v)$, respectively.

3. The elimination principle discussed below can be applied without concern over A's reducibility; however, if A is known to be irreducible, fewer solutions need to be tested.

4. On invertibility, see Basset, Maybee, and Quirk (1968). On necessarily determinable entries, see Lady (1983). Lady provided necessary and sufficient conditions for the invertibility of $\operatorname{sgn} A$ put into SF1, but not (necessarily) put into SF2. The algorithmic burden of detecting Lady's conditions is severe compared to putting A into SF2, and then detecting the conditions given in Category II.1. Alternatively, once A is shown to be a QI-matrix, Lady's approach enables the immediate determination of the signs of the elements of A^{-1} incident on the transpose of A's nonzero elements prior to putting A into SF2, without having to account for the sign changes in A's columns necessary to put A into SF2. On possibly determinable entries, see Maybee and Quirk (1969). On entries of $\operatorname{sgn} A^{-1}$ that cannot be determined, see Lady and Maybee (1983).

5. All possible solutions are formed by listing the base 2 number expressions of the solution number from 0 to $2^n - 1$ (for irreducible systems for which zeros can be ignored). The kth entry of the solution sign pattern sgn x is then given as the kth digit in the base 2 number with 0 replaced by -1 (a negative sign).

6. For example, consumer preferences for static, certain choices are representable by an ordinal measure, preferences under conditions of risk are represented by an interval measure, and intertemporal preferences are sometimes represented by a ratio or fully quantitative measure.

7. The Jacobian matrices corresponding to actual models are usually very sparse. As a result, instances in which a Jacobian matrix is a QI-matrix may not be as unlikely as the stringent conditions for invertibility suggest. For example, as shown in Chapter 4 the 19×19, irreducible Jacobian matrix corresponding to a model of oil pricing and trade flows is a QI-matrix.

8. Note that the analysis of ord A is intended to be applicable to any sign- and rank-preserving transformation of its entries; e.g., for this matrix such matrices are all matrices with all negative entries ranked as above.

9. Main diagonal elements themselves can be considered to correspond to "one-cycles," representing inference in a reflexive sense. Given this, the correspondence of collections of disjoint cycles to nonzero terms in det(A) is exhaustive.

10. The conditions could be simplified somewhat by transforming A such that for A an RI-matrix, if det $A \neq 0$, then sgn det $A = (-1)^n$. Such a transformation would be accomplished if, for $n(UFT) > n(FT)$, the matrix were reindexed to bring an unfriendly term onto the main diagonal, and then the columns were multiplied by -1 as necessary such that $a_{ii} < 0$ for all i. Now the signs of all of the nonzero terms in the expansion of det A have been reversed. This is eventually proposed in the analysis of cycle dominant matrices. When put into this form, matrices are defined as being in SF3*.

11. We use "nord" for "normalized ordinal."

12. McKenzie (1960a, 48). The definition has been particularized here for the case of an irreducible matrix.

13. See Chapter 8 for a comprehensive discussion and interpretation of stable matrices.

Chapter 4

1. Lady (1993) developed SGNSOLVE.EXE and PREPARE.EXE to assist in assessing a matrix's structural and qualitative characteristics.

2. The computer program CAUSOR was developed to assist in identifying and working with complex inference structures that might be present in large models (Gilli 1992).

3. *Oil Market Simulation User's Manual*, DOE/EIA-M028(90) (Washington, D.C.: Energy Information Administration [EIA], 1990). Earlier work presenting the mathematical basis for OMS and the outcome of estimating its embodied coefficients is given in *Final Report, The Oil Market Simulation Model, Model Documentation Report*, DOE/EI/19656-2, prepared by System Sciences, Inc. (Washington, D.C.: EIA, 1985). The qualitative analysis of OMS's Jacobian

matrix (Lady 1993) was originally done as part of EIA's quality assurance program, and conveyed to the EIA under Contract PN 0890-055-07-L01 under agreement with ANSTEC, Inc., Washington, D.C. The results of this analysis are also presented in Hale and Lady (1995).

4. The analysis summarized in this section was conducted for EIA's Quality Assurance Division and reported in G. Lady, *Structural and Qualitative Analysis of the Onshore Lower 48 Conventional Oil and Gas Supply Model* (conveyed to the EIA under Contract 91EI21938, Task Assignment 93023 under agreement with ANSTEC, Inc., Washington, D.C., 1994). The description of the Oil and Gas Supply Model on which the analysis is based is presented in *Documentation of the Oil and Gas Supply Model (OGSM)*, prepared by the Oil and Gas Analysis Branch, Office of Integrated Analysis and Forecasting, EIA (Washington, D.C., January 7, 1994).

5. SUCDCFON(I = 1) and DRYDCFON(I = 1) appear at this point as parameters (i.e., they have not been defined as variables, as have the corresponding magnitudes for I = 2).

6. Notation problem: parameter 26 is also Q(initial resource estimate). The production rate is from the Petroleum Market Model (PMM).

7. This ratio (PR) must be smaller than 1/2. Hence, as it increases, $PR(1 - PR)$ must also increase.

8. Klein 1950, as presented in Zellner and Theil 1962. Brouwer et al. (1989) analyze this model using its signed directed graph; however, there is an error in transforming the matrix to a standard form prior to constructing the graph.

Chapter 8

1. Since no *numeraire* is chosen, the equilibrium price vector is only unique up to a (positive) scalar multiple. Arrow and Hurwicz (1958) prove global stability for the system with *numeraire*.

2. Since every matrix that is Metzler is stable if and only if it is Hicksian, this means that any principal minor of such a matrix can be chosen arbitrarily large in absolute value relative to other principal minors, with stability being preserved, as long as the matrix remains Hicksian. Thus, in any Routh-Hurwitz determinantal condition, after cancellation of like terms with opposite signs, every remaining nonzero term involving the product of principal minors with principal minors has to be positive. The only remaining terms in any determinantal conditions are those that arise because of the cancellation process. Such terms will involve the product of a (positive) cycle with a principal minor or a product of principal minors. The only time that such a term will arise is when the principal minor in which the positive cycle is located has the incorrect (−) sign within the determinantal condition. Since in a Hicksian matrix, positive cycles have signs opposite to that of the principal minor in which they are located, this means that after cancellations, terms in which positive cycles appear are always positive.

Bibliography

Albin, Peter, and Hans Gottinger, 1983, "Structure and Complexity in Economic and Social Systems," *Mathematical Social Sciences*, Vol. 5, 253–268.

Allingham, Michael, and Michio Morishima, 1970, Some Extensions of the Gross Substitute System, University of Essex. Duplicated.

———, 1973, "Qualitative Economics and Comparative Statics," in M. Morishima, ed., *Theory of Demand*: *Real and Monetary*, Oxford: Oxford University Press.

Anderson, Sabra, 1971, *Graph Theory and Finite Combinatorics*, Chicago: Mark Publishing.

Archibald, G. C., 1965, "The Qualitative Content of Maximizing Models," *Journal of Political Economy*, Vol. 75, 27–36.

Arrow, Kenneth, 1959, "Toward a Theory of Price Adjustment," in Moses Abramovitz, ed., *The Allocation of Resources*, Stanford: Stanford University Press, 41–51.

———, 1974, "Stability Independent of Adjustment Speed," in George Horwich and Paul Samuelson, eds., *Trade, Stability and Macroeconomics*, New York: Academic Press, 181–201.

Arrow, Kenneth, and G. Debreu, 1954, "Existence of an Equilibrium for a Competitive Economy," *Econometrica*, Vol. 22, 265–290.

Arrow, Kenneth, and A. Enthoven, 1956, "A Theorem on Expectations and the Stability of Equilibrium," *Econometrica*, Vol. 24, 288–293.

Arrow, Kenneth, and Frank Hahn, 1971, *General Competitive Analysis*, San Francisco: Holden-Day.

Arrow, Kenneth, and Leonid Hurwicz, 1958, "On the Stability of the Competitive Equilibrium I," *Econometrica*, Vol. 26, 522–552.

———, 1959, "On the Stability of the Competitive Equilibrium II," *Econometrica*, Vol. 27, 82–109.

———, 1960a, "Some Remarks on the Equilibria of Economic Systems," *Econometrica*, Vol. 28, 640–646.

———, 1960b, "Competitive Stability under Weak Gross Substitutability: The 'Euclidean Distance' Approach," *International Economic Review*, Vol. 1, 38–49.

Arrow, Kenneth, and Maurice McManus, 1958, "A Note on Dynamic Stability," *Econometrica*, Vol. 26, 448–454.

Barrett, Wayne, and Charles Johnson, 1984, "Determinantal Formulae for Matrices with Sparse Inverses," *Linear Algebra and Its Applications*, Vol. 56, 73–88.

Barrett, Wayne, Charles Johnson, D. D. Olesky, and Paula van den Driessche, 1987, "Inherited Matrix Entries: Principal Submatrices of the Inverse," *SIAM Journal of Algebra and Discrete Methods*, Vol. 8, 313–322.

Bassett, Lowell, Hamid Habibagahi, and James Quirk, 1967, "Qualitative Economics and Morishima Matrices," *Econometrica*, Vol. 35, 221–233.

Bassett, Lowell, John Maybee, and James Quirk, 1968, "Qualitative Economics and the Scope of the Correspondence Principle," *Econometrica*, Vol. 36, 544–563.

Berge, C., 1962, *Graph Theory and Its Applications*, New York: Wiley.

Berman, A., and D. Hershkowitz, 1985, "Graph Theoretic Methods in Studying Stability," *Contemporary Mathematics*, Vol. 47, 1–6.

Berman, A., and R. Plemmons, 1979, *Nonnegative Matrices in the Mathematical Sciences*, New York: Academic Press.

Berndsen, Ron, and Hennie Daniels, 1990, "Qualitative Dynamics and Causality in a Keynesian Model," *Journal of Dynamic Economics and Control*, Vol. 14, 435–450.

Berndt, Ernest R., 1991, *The Practice of Econometrics, Classic and Contemporary*, New York: Addison Wesley.

Blad, M., 1978, "On the Speed of Adjustment in the Classical Tatonnement Process: A Limit Result," *Journal of Economic Theory*, Vol. 19, 186–191.

Blanchard, O., and M. Plantes, 1977, "A Note on Gross Substitutability of Financial Assets," *Econometrica*, Vol. 45, 769–771.

Bobrow, D., ed., 1985, *Qualitative Reasoning about Physical Systems* Cambridge: MIT Press.

Bone, T., Clark Jeffries, and Victor Klee, 1988, "A Qualitative Analysis of $x' = Ax + b$," *Discrete Applied Mathematics*, Vol. 20, 9–30.

Bourgine, Paul, 1990, "A Logical Approach to Problem Representation," *Journal of Economic Dynamics and Control*, Vol. 14, 451–464.

Brouwer, F., J. Maybee, P. Nijkamp, and H. Voogd, 1989, "Sign Solvability in Economic Models through Plausible Restrictions," *Atlantic Economic Journal*, 17, 21–26.

Brouwer, F., and P. Nijkamp, 1984, "Qualitative Structure Analysis of Complex Systems," in P. Nijkamp, H. Leitner, and N. Wrigley, eds., *Measuring the Unmeasurable*, Hague: Martnurs Nijhoff.

Brualdi, R., and B. Shader, 1991, "On Sign Nonsingular Matrices and the Conversion of the Permanent into the Determinant," in Gitzmann and Sturmfels, eds., *Applied Geometry and Discrete Mathematics*, Victor Klee Festschrift, 117–134.

Bushaw, D., and R. Clower, 1957, *Introduction to Mathematical Economics*, Homewood, Ill.: Richard D. Irwin.

Carlson, David, 1967, "Weakly Sign Symmetric Matrices and Some Determinantal Inequalities," *Colloquium Mathematicum*, Vol. 17, 123–129.

———, 1974, "A Class of Stable Matrices," *Journal of Research of the National Bureau of Standards, Section B*, Vol. 78B, 1–2.

———, 1988, "Nonsingularity Criteria for Matrices Involving Combinatorial Considerations," *Linear Algebra and Its Applications*, Vol. 107, 41–56.

Carlson, David, B. N. Datta, and Charles Johnson, 1982, "A Semi-Definite Lyapunov Theorem and the Characterization of Tridiagonal D-Stable Matrices," *SIAM Journal of Algebra and Discrete Methods*, Vol. 3, 293–304.

Chitre, V., 1974, "A Note on the Three Hicksian Laws of Comparative Statics for the Gross Substitute Case," *Journal of Economic Theory*, Vol. 8, 397–400.

Clark, B. L., 1975a, "Theorems on Chemical Network Stability," *Journal of Chemical Physics*, Vol. 62, 773–775.

———, 1975b, "Stability of Topologically Similar Chemical Reactions," *Journal of Chemical Physics*, Vol. 62, 3726–3738.

Debreu, Gerard, 1959, *Theory of Value*, New York: Wiley.

Debreu, Gerard, and Herstein, I., 1953, "Nonnegative Square Matrices," *Econometrica*, Vol. 21, 597–607.

Don, F., and J. Henk, 1990, "Some Issues in Solving Large Sparse Systems of Equations," *Journal of Economic Dynamics and Control*, Vol. 14, 313–325.

Energy Information Administration (EIA), 1990, *Oil Market Simulation User's Manual*, DOE/EIA-M028(90), Washington, D.C.

Eschenback, Carolyn, and Charles Johnson, 1991, "Sign Patterns That Require Real, Nonreal, or Pure Imaginary Eigenvalues," *Linear and Multilinear Algebra*, Vol. 29, 299–311.

Farley, Arthur, and Kuan-Pin Lin, 1990, "Qualitative Reasoning in Economics," *Journal of Economic Dynamics and Control*, Vol. 14, 465–490.

Fiedler, Miroslav, and Vlastimil Ptak, 1966, "Some Generalizations of Positive Definiteness and Stability," *Numerische Mathematik*, Vol. 9, 163–172.

Fisher, Frank, 1972, "Gross Substitutes and the Utility Function," *Journal of Economic Theory*, Vol. 4, 82–87.

Fontaine, Garbely, and M. H. Gilli, 1991, "Qualitative Solvability in Economic Models," *Computer Science in Economics and Management*, Vol. 5, 539–548.

Fox, George, and James Quirk, 1985, Uncertainty and Input-Output Analysis, Economics Group, Jet Propulsion Laboratory, Paper No. 23. Duplicated.

Frobenius, G., 1908, "Uber Matrizen aus positiven Elementen," *Sitzungsberichte der koniglich preussischen Akadamie der Wissenschaften*, 471–476.

Fuller, A., and M. Fisher, 1958, "On the Stabilization of Matrices and the Convergence of the Linear Iterative Process," *Proceedings of the Cambridge Philosophical Society*, Vol. 54, 417–425.

Gale, David, 1969, *On the Theory of Linear Economic Models*, New York: McGraw-Hill.

Gantmackher, F. R., 1959, *The Theory of Matrices*, New York: Chelsea Pub. Co.

Garbely, A., and M. H. Gilli, 1991, "Qualitative Decomposition of the Eigenvalue Problem in a Dynamic System," *Journal of Economic Dynamics and Control*, Vol. 15, 539–548.

Garcia, G., 1972, "Olech's Theorem and the Dynamic Stability of Theories of the Rate of Interest," *Journal of Economic Theory*, Vol. 4, 541–544.

Gillen, W. J., and A. Guccione, 1990, "The Introduction of Constraints in Lancaster's Qualitative Comparative Statics Algorithm," *Quarterly Journal of Economics*, Vol. 105, 1053–1061.

Gilli, M. H., 1992, computer code listing for Analysis of Causal Structures Version 1.00, Department of Econometrics, University of Geneva (8384 lines).

Gorman, Terrance, 1964, "More Scope for Qualitative Economics," *Review of Economic Studies*, Vol. 31, 65–68.

Greenberg, Harvey, 1981, "Measuring Complementarity and Qualitative Determinacy in Matricial Forms," in Harvey Greenberg and John Maybee, eds. *Computer-Assisted Analysis and Model Simplification*, New York: Academic Press, 497–522.

Greenberg, Harvey, J. Lundgren, and John Maybee, 1981a, "Graph Theoretic Methods for the Qualitative Analysis of Rectangular Matrices," *SIAM Journal of Algebra and Discrete Methods*, Vol. 2, 227–239.

———, 1981b, "Graph-Theoretic Foundations of Computer-Assisted Analysis," in Harvey Greenberg and John Maybee, eds. *Computer-Assisted Analysis and Model Simplification*, New York: Academic Press, 481–495.

Greenberg, Harvey, and John Maybee, eds., 1981, *Computer-Assisted Analysis and Model Simplification*, New York: Academic Press.

Habibigahi, Hamid, 1965, Walrasian Stability: Qualitative Economics, Purdue University, Krannert Institute Series No. 145.

———, 1966, Qualitative Analysis of Dynamic Stability in General Equilibrium Theory, unpublished thesis, Purdue University.

Habibagahi, Hamid, and James Quirk, 1973, "Hicksian Stability and Walras' Law," *Review of Economic Studies*, Vol. 40, 249–258.

———, 1986, Qualitative Analysis of the Competitive Equilibrium, Economics Group, Jet Propulsion Laboratory, Paper No. 25. Duplicated.

Hahn, Frank, 1958, "Gross Substitutes and the Dynamic Stability of General Equilibrium," *Econometrica*, Vol. 26, 169–170.

Hale, D., and G. Lady, 1995, "Qualitative Comparative Statics and Audits of Model Performance," *Linear Algebra and Its Applications*, Vol. 217, 141–154.

Hansen, P., 1983, "Recognizing Sign Solvable Graphs," *Discrete Applied Mathematics*, Vol. 6, 237–241.

———, 1984, "Shortest Paths in Signed Graphs," in R. E. Burkard, R. A. Cunninghame-Gren, and U. Zimmermann, eds., *Algebraic and Combinatorial Methods in Operations Research*, Amsterdam: North-Holland, 201–214.

Harary, Frank, 1981, "Structural Models and Graph Theory," in Harvey Greenberg and John Maybee, eds., *Computer-Assisted Analysis and Model Simplification*, New York: Academic Press, 31–58.

Harary, F., R. Norman, and D. Cartwright, 1965, *Structured Models: An Introduction to the Theory of Directed Graphs*, New York: Wiley.

Hawkins, D., and Herbert Simon, 1949, "Note: Some Conditions of Macroeconomic Stability," *Econometrica*, Vol. 17, 245–248.

Hicks, John R., 1939, *Value and Capital*, Oxford: Oxford University Press.

Hotson, John, and Hamid Habibagahi, 1972, "Comparative Static Analysis of Harrod's Dichotomy," *Kyklos*, Vol. 25, 326–344.

Householder, A., 1964, *The Theory of Matrices in Numerical Analysis*, New York: Blausdell Pub. Co.

Hurwitz, A., 1895, "Uber die Bedingungen, unter welchen eine Gleichung nur Wurzeln mit negativen reellen Teilen Besitzt," *Mathematische Annalen*, Vol. 46, 273–284.

Ichioka, Osamu, 1979, "The Local Stability of the Morishima Case," *Economic Studies Quarterly*, Vol. 30, 83–86.

Ishida, Y., N. Adachi, and H. Tokumaru, 1981, "Some Results on the Qualitative Theory of Matrices," *Transactions of Social Instruments and Control Engineering*, Vol. 17, 49–55.

Jeffries, Clark, 1974, "Qualitative Stability and Digraphs in Model Ecosystems," *Ecology*, Vol. 55, 1415–1419.

———, 1986, "Qualitative Stability of Certain Nonlinear Systems," *Linear Algebra and Its Applications*, Vol. 75, 133–144.

Jeffries, Clark, and Charles Johnson, 1988, "Some Sign Patterns That Preclude Matrix Stability," *SIAM Journal of Matrix Analysis and Applications*, Vol. 9, 19–25.

Jeffries, Clark, Victor Klee, and Pauline van den Driessche, 1977, "When Is a Matrix Sign Stable?" *Canadian Journal of Mathematics*, Vol. 29, 315–326.

———, 1987, "Qualitative Stability of Linear Systems," *Linear Algebra and Its Applications*, Vol. 87, 1–48.

Jeffries, Clark, and Pauline van den Driessche, 1988, "Eigenvalues of Matrices with Tree Graphs," *Linear Algebra and Its Applications*, Vol. 101, 109–120.

———, 1991, "Qualitative Stability and Solvability of Difference Equations," *Linear and Multilinear Algebra*, Vol. 30, 275–282.

Johnson, Charles, 1974a, "Sufficient Conditions for D-Stability," *Journal of Economic Theory*, Vol. 9, 53–62.

———, 1974b, "Second, Third and Fourth Order D-Stability," *Journal of Research of the National Bureau of Standards*, Vol. 78B, 11–14.

———, 1988, "Combinatorial Matrix Analysis: An Overview," *Linear Algebra and Its Applications*, Vol. 107, 3–15.

Johnson, Charles, F. Uhlig, and D. Warner, 1982, "Sign Patterns, Nonsingularity, and Solvability of $Ax = b$," *Linear Algebra and Its Applications*, Vol. 47, 1–9.

Johnson, Charles, and Paula van den Driessche, 1988, " Interpolation of D-Stability and Sign Stability," *Linear and Multilinear Algebra*, Vol. 23, 363–368.

Kamien, Morton, 1964, "A Note on Complementarity and Substitution," *International Economic Review*, Vol. 5, 221–227.

Kennedy, C., 1970, "The Stability of the Morishima System," *Review of Economic Studies*, Vol. 37, 173–175.

Keynes, J. M., [1936] 1960, *The General Theory of Employment Interest and Money*, reprint, New York: Harcourt, Brace.

Khalil, Hassan, 1980, "A New Test for D-Stability," *Journal of Economic Theory*, Vol. 23, 120–122.

Klee, Victor, 1987, "Recursive Structure of S-Matrices and an O (m-squared) Algorithm for Recognizing Strong Sign Solvability," *Linear Algebra and Its Applications*, Vol. 96, 233–247.

———, 1989, "Sign-Patterns and Stability," in Fred Roberts, ed., *Applications of Combinatorics and Graph Theory to the Biological and Social Sciences*, New York: Springer Verlag, 203–219.

Klee, Victor, and R. Ladner, 1981, "Qualitative Matrices: Strong Sign-Solvability and Weak Satisfiability," in Harvey Greenberg and John Maybee, eds., *Computer-Assisted Analysis and Model Simplification*, New York: Academic Press, 293–323.

Klee, Victor, R. Ladner, and R. Manber, 1984, "Signsolvability Visited," *Linear Algebra and Its Applications*, Vol. 59, 181–197.

Klee, Victor, and Paula van den Driessche, 1977, "Linear Algorithms for Testing the Sign Stability of a Matrix and Finding Z-maximum Matchings in Acyclic Graphs," *Numerical Mathematics*, Vol. 28, 273–285.

Klein, D., 1982, "Tree Diagonal Matrices and Their Inverses," *Linear Algebra and Its Applications*, Vol. 42, 109–117.

Klein, L., 1950, *Economic Fluctuations in the U.S., 1921–1941*, Cowles Commission for Research in Economics Monograph No. 11, New York: Wiley.

Kupiers, B., 1986, "Qualitative Simulation," *Artificial Intelligence*, Vol. 29, 289–338.

Lady, George, 1981, "Organizing Analysis," in Harvey Greenberg and John Maybee, eds., *Computer-Assisted Analysis and Model Simplification*, New York: Academic Press, 1–17.

———, 1983, "The Structure of Qualitatively Determinate Relationships," *Econometrica*, Vol. 51, 197–218.

———, 1993, SGNSOLVE.EXE and PREPARE.EXE Analysis Packages, available from Energy Information Administration, U.S. Department of Energy, Washington, D.C.

———, 1995, "Robust Economic Models," *Journal of Economic Dynamics and Control*, Vol. 19, 481–501.

———, 1996, "Detecting Stable Matrices," *Annals of Mathematics and Artificial Intelligence*, Vol. 17, 29–36.

Lady, George, Thomas Lundy, and John Maybee, 1995, "Nearly Sign-Nonsingular Matrices," *Linear Algebra and Its Applications*, Vol. 220, 229–248.

Lady, George, and John Maybee, 1983, "Qualitatively Invertible Matrices," *Mathematical Social Sciences*, Vol. 6, 397–407.

Lancaster, Kelvin, 1962, "The Scope of Qualitative Economics," *Review of Economic Studies*, Vol. 29, 99–132.

———, 1964, "Partitionable Systems and Qualitative Economics," *Review of Economic Studies*, Vol. 31, 69–72.

———, 1965, "The Theory of Qualitative Linear Systems," *Econometrica*, Vol. 33, 395–408.

———, 1966, "The Solution of Qualitative Comparative Statics Problems," *Quarterly Journal of Economics*, Vol. 53, 278–295.

Lange, Osca, 1944, *Price Flexibility and Employment*, Cowles Foundation Monograph No. 8, Bloomington, Ill.: Principia Press.

Larson, Bruce, 1980, "An Algebraic Approach to Qualitative Knowledge," *Quality and Quantity*, Vol. 14, 355–362.

Levins, R., 1974a, "The Qualitative Analysis of Partially Specified Systems," *Annals of the New York Academy of Science*, Vol. 231, 123–138.

———, 1974b, "Problems of Signed Digraphs in Ecological Theory," in S. Levin, ed., *Ecosystem Analysis and Prediction*, New York: Wiley, 264–276.

Logofet, D., 1987, "On the Hierarchy of Subsets of Stable Matrices," *Soviet Math. Dokl.*, Vol. 34, 247–250.

Logofet, D., and N. Ulyanov, 1982a, "Necessary and Sufficient Conditions for Sign Stability of Matrices," *Soviet Math. Dokl.*, Vol. 25, 676–680.

———, 1982b, "Sign Stability in Model Ecosystems: A Complete Class of Sign-Stable Patterns," *Ecological Modelling*, Vol. 16, 173–189.

Lundgren, J. Richard, and John Maybee, 1984, "A Class of Maximal *L*-Matrices," *Congressus Numerantum*, Vol. 44, 239–249.

Lundy, Thomas, and John Maybee, 1991a, Zero Submatrices and Matrix and Digraph Complexity, University of Colorado. Duplicated.

———, 1991b, Inverses of Sign Nonsingular Matrices, University of Colorado. Duplicated.

Manber, Rachel, 1982, "Graph-Theoretical Approach to Qualitative Solvability of Linear Systems," *Linear Algebra and Its Applications*, Vol. 48, 457–470.

May, R., 1973, "Qualitative Stability in Model Ecosystems," *Ecology*, Vol. 54, 638–641.

Maybee, John, 1966a, Remarks on the Theory of Cycles in Matrices, Mathematics Department, Purdue University. Duplicated.

———, 1966b, "New Generalizations of Jacobi Matrices," *SIAM Journal of Applied Mathematics*, Vol. 14, 1032–1037.

———, 1967, "Matrices of Class J2," *Journal of Research of National Bureau of Standards, Section B*, Vol. 71, 215–224.

———, 1974, "Combinatorially Symmetric Matrices," *Linear Algebra and Its Applications*, Vol. 8, 529–537.

———, 1980, "Sign Solvable Graphs," *Discrete Applications in Mathematics*, Vol. 2, 57–63.

———, 1981, "Sign Solvability," in Harvey Greenberg and John Maybee, eds., *Computer-Assisted Analysis and Model Simplification*, New York: Academic Press, 201–259.

———, 1988, "Some Possible New Directions for Combinatorial Matrix Analysis," *Linear Algebra and Its Applications*, Vol. 107, 23–40.

———, 1989, "Qualitatively Stable Matrices and Convergent Matrices," in Fred Robert, ed., *Applications of Combinatorics and Graph Theory to the Biological and Social Sciences*, New York: Springer Verlag, 245–258.

Maybee, John, and Stuart Maybee, 1984, "An Algorithm for Identifying Morishima and Anti-Morishima Matrices and Balanced Digraphs," *Mathematical Social Sciences*, Vol. 6, 99–103.

Maybee, John, D. Olesky, Pauline van den Driessche, and G. Weiner, 1989, "Matrices, Digraphs, and Determinants," *SIAM Journal of Matrix Analysis and Applications*, Vol. 10, 500–519.

Maybee, John, and James Quirk, 1969, "Qualitative Problems in Matrix Theory," *SIAM Review*, Vol. 11, 30–51.

———, 1973, The GM Matrix Problem, Department of Computer Sciences, University of Colorado. Duplicated.

Maybee, John, and Daniel Richman, 1988, "Some Properties of GM-matrices and Their Inverses," *Linear Algebra and Its Applications*, Vol. 107, 219–236.

Maybee, John, and H. Voogd, 1984, "Qualitative Impact Analysis through Sign-Stability: A Review," *Environmental Planning Bulletin*, Vol. 11, 365–374.

Maybee, John, and Gerry Weiner, 1987, "*L*-Functions and Their Inverses," *SIAM Journal of Algebra and Discrete Methods*, Vol. 8, 67–76.

———, 1988, "From Qualitative Matrices to Quantitative Restrictions," *Linear and Multilinear Algebra*, Vol. 22, 229–248.

Maybee, Stuart, 1986, *A Method for Identifying Sign Solvable Systems*, M.S. thesis, University of Colorado.

McDuffee, C., 1946, *The Theory of Matrices*, New York: Chelsea Pub. Co.

McKenzie, Lionel, 1960a, "The Matrix with Dominant Diagonal and Economic Theory," in K. Arrow, S. Carlin, and H. Scarf, eds., *Proceedings of a Symposium in Mathematical Methods in the Social Sciences*, Stanford: Stanford University Press.

———, 1960b, "Stability of Equilibrium and the Value of Positive Excess Demand," *Econometrica*, Vol. 28, 606–617.

McManus, Maurice, 1958, Stability of the Gross Substitute Case, Stanford. Duplicated.

Metzlar, A., 1989, Minimum Transversal of Cycles in Intercyclic Digraphs, Ph.D. diss., University of Waterloo (Canada).

Metzler, Lloyd, 1945, "Stability of Multiple Markets: The Hicks Conditions," *Econometrica*, Vol. 13, 277–292.

Michel, A. N., and R. K. Miller, 1987, *Qualitative Analysis of Large Scale Systems*, New York: Academic Press.

Milgrom, Paul, 1994, "Comparing Optima: Do Simplifying Assumptions Affect Conclusions?" *Journal of Political Economy*, Vol. 102, 607–615.

Milgrom, Paul, and John Roberts, 1990, "Rationalization, Learning and Equilibrium in Games with Strategic Complementarities," *Econometrica*, Vol. 58, 1255–1278.

———, 1994, "Comparing Equilibria," *American Economic Review*, Vol. 84, 441–459.

Milgrom, Paul, and Chris Shannon, 1994, "Monotone Comparative Statics," *Econometrica*, Vol. 62, 157–180.

Morishima, Michio, 1952, "On the Laws of Change of the Price System in an Economy Which Contains Complementary Commodities," *Osaka Economic Papers*, Vol. 1, 101–113.

———, 1957, "Notes on the Theory of the Stability of Multiple Exchange," *Review of Economic Studies*, Vol. 24, 177–186.

———, 1964, *Equilibrium, Stability, and Growth*, New York: Oxford University Press.

———, 1970, "A Generalization of the Gross Substitute System," *Review of Economic Studies*, Vol. 37, 177–186.

———, 1973, *Theory of Demand, Real and Monetary*, Oxford: Oxford University Press.

Mosak, J., 1944, *General Equilibrium Theory in International Trade*, Cowles Foundation Monograph No. 7. Bloomington, Ill.: Principia Press.

Mosenson, R., and E. Dror, 1972, "A Solution to the Qualitative Substitution Problem in Demand Theory," *Review of Economic Studies*, Vol. 39, 433–442.

Mukherji, Anjan, 1972, "On Complementarity and Stability," *Journal of Economic Theory*, Vol. 4, 442–457.

———, 1973a, On the Laws of Comparative Statics and the Correspondence Principle, Jaharal Nehru University. Duplicated.

———, 1973b, "On the Sensitivity of Stability Results to the Choice of the Numeraire," *Review of Economic Studies*, Vol. 40, 427–433.

Murota, K., and M. Iri, 1985, "Structural Solvability of Systems of Equations," *Japanese Journal of Applied Mathematics*, Vol. 1, 247–271.

Negishi, T., 1958, "A Note on the Stability of an Economy Where All Goods Are Gross Substitutes," *Econometrica*, Vol. 26, 445–447.

———, 1962, "The Stability of a Competitive Economy: A Survey Article," *Econometrica*, Vol. 30, 635–669.

Newman, P. K., 1959, "Some Notes on Stability Conditions," *Review of Economic Studies*, Vol. 27, 1–9.

———, 1961, "Approaches to Stability Analysis," *Economica*, Vol. 28, 12–29.

Nikaido, H., 1964, "Generalized Gross Substitutability and Extremization," in M. Dresher, L. S. Shapley, and A. W. Tucker, eds., *Advances in Game Theory*, Princeton: Princeton University Press, 55–68.

———, 1968, *Convex Structures and Economic Theory*, New York: Academic Press.

OGSM, 1994. See note 4 to Chapter 4.

Ohyama, M., 1972, "On the Stability of Generalized Metzlerian Systems," *Review of Economic Studies*, Vol. 39, 193–204.

Olech, C., 1963, "On the Global Stability of an Autonomous System on the Plane," *Contributions to Differential Equations*, Vol. 1, 389–400.

Patinkin, D., 1952, "The Limitations to Samuelson's 'Correspondence Principle,'" *Metroeconomica*, Vol. 4, 37–43.

Pearce, I., 1974, "Matrices with Dominating Diagonal Blocks," *Journal of Economic Theory*, Vol. 9, 159–170.

Perron, O., 1907, "Zur Theorie der Matrices," *Mathematische Annalen*, Vol. 64, 248–263.

Poole, George, and Thomas Bouillion, 1974, "A Survey on M-matrices," *SIAM Review*, Vol. 16, 419–427.

Popper, K., [1934] 1959, *The Logic of Scientific Discovery*, reprint, New York: Harper and Row.

Provan, J. Scott, 1983, "Determinacy in Linear Systems and Networks," *SIAM Journal of Algebra and Discrete Methods*, Vol. 4, 262–278.

Quirk, James, 1968a, "Comparative Statics under Walras's Law: The Case of Strong Dependence," *Review of Economic Studies*, Vol. 35, 11–21.

———, 1968b, "The Correspondence Principle: A Macroeconomic Application," *International Economic Review*, Vol. 9, 294–306.

———, 1969, "The Competitive Equilibrium: A Qualitative Analysis," in Martin Beckmann and H. Kunzi, eds., *Economic Models, Estimation and Risk Programming*, Berlin: Springer Verlag.

———, 1970, "Complementarity and Stability of the Competitive Equilibrium," *American Economic Review*, Vol. 60, 358–363.

———, 1972a, Purely Qualitative Models, Caltech. Duplicated.

———, 1972b, Qualitative Analysis of the Competitive Equilibrium, Caltech. Duplicated.

———, 1974, "A Class of Generalized Metzlerian Matrices," in George Horwich and Paul Samuelson, eds., *Trade, Stability, and Macroeconomics*, New York: Academic Press, 203–219.

———, 1981, "Qualitative Stability of Matrices and Economic Theory: A Survey Article," in Harvey Greenberg and John Maybee, eds., *Computer-Assisted Analysis and Model Simplification*, New York: Academic Press, 113–164.

———, 1986, "Qualitative Economics," in *The New Palgrave Dictionary of Economics*, New York: Macmillan.

———, 1992, *Qualitative Analysis of Economic Models, Qualitative Matrices, and Related Matters: A Survey of the Literature*, available from Energy Information Administration, U.S. Department of Energy, Washington, D.C.

———, 1997, "Qualitative Comparative Statics," *Journal of Mathematical Economics*, Vol. 28, 127–154.

Quirk, James, and Richard Ruppert, 1965, "Qualitative Economics and the Stability of Equilibrium," *Review of Economic Studies*, Vol. 32, 311–326.

———, 1967, Global Stability and Phase Diagrams, Paper No. 5, Department of Economics, University of Kansas. Duplicated.

———, 1968, "Maximization and the Qualitative Calculus," in James Quirk and Arvid Zarley, eds., *Papers in Quantitative Economics*, Lawrence: University of Kansas Press, 73–101.

Quirk, James, and Rubin Saposnik, 1968, *Introduction to General Equilibrium Theory and Welfare Economics*, New York: McGraw-Hill.

Rader, Trout, 1968, "Normally, Factor Inputs Are Never Gross Substitutes," *Journal of Political Economy*, Vol. 78, 38–43.

———, 1972, "Impossibility of Qualitative Economics: Excessively Strong Correspondence Principles in Production-Exchange Economies," *Zeitschrift fur Nationalokonomie*, Vol. 32, 397–416.

Redheffer, R., and W. Walter, 1984, "Solution of the Stability Problem for a Class of Generalized Prey-Predator Systems, *Journal of Differential Equations*, Vol. 55, 245–263.

Redheffer, R., and Z. Zhou, 1982, "A Class of Matrices Connected with Volterra Prey-Predator Equations," *SIAM Journal of Algebra and Discrete Methods*, Vol. 3, 122–134.

Ritschard, Gilbert, 1983, "Computable Qualitative Comparative Static Techniques," *Econometrica*, 51, 1145–1168.

Roberts, Fred, 1971, "Signed Digraphs and the Growing Demand for Energy," *Environment and Planning*, Vol. 3, 395–410.

———, 1981, "Structural Models and Graph Theory," in Harvey Greenberg and John Maybee, eds., *Computer-Assisted Analysis and Model Simplification*, New York: Academic Press, 59–67.

———, 1989, "Applications of Combinatorics and Graph Theory to the Biological and Social Sciences: Seven Fundamental Ideas," in Fred Roberts, ed., *Applications of Combinatorics and Graph Theory to the Biological and Social Sciences*, New York: Springer Verlag, 1989, 1–37.

Rosenblatt, D., 1957, "On Linear Models and the Graphs of Minkowski-Leontief Matrices," *Econometrica*, Vol. 25, 323–338.

Routh, E. J., 1877, *A Treatise on the Stability of a Given State of Motion*, London: Macmillan.

Samuelson, Paul, 1941, "The Stability of Equilibrium: Comparative Statics and Dynamics," *Econometrica*, Vol. 9, 97–120.

———, 1944, "The Relation between Hicksian Stability and True Dynamic Stability," *Econometrica*, Vol. 12, 256–257.

———, 1947, *Foundations of Economic Analysis*, Cambridge: Harvard University Press.

Sato, R., 1972, "The Stability of the Competitive System Which Contains Complementary Goods," *Review of Economic Studies*, Vol. 39, 495–499.

———, 1973, "On the Stability Properties of Dynamic Economic Systems," *International Economic Review*, Vol. 14, 753–764.

Sato, R., and T. Koizumi, 1970, "Substitutability, Complementarity, and the Theory of Derived Demand," *Review of Economic Studies*, Vol. 37, 107–118.

Sattinger, M., 1975, "Local Stability When Initial Holdings Are Near Equilibrium Holdings," *Journal of Economic Theory*, Vol. 11, 161–167.

Shields, R. W., and J. B. Pearson, 1976, "Structual Controllability of Multi-input Linear Systems," *IEEE Transactions on Automatic Control*, AC-21, 203–212.

Shirakura, T., 1986, "Jeffries Colour Point Method and the Simons-Homan Model" (in Japanese), *Sociological Theory and Methods*, Vol. 1, 57–70.

Smithies, A., 1942, "The Stability of Competitive Equilibrium," *Economica*, Vol. 10, 258–274.

Solimano, F., and E. Beretta, 1982, "Graph Theoretical Criteria for Stability and Boundedness of Predator-Prey Systems," *Bulletin of Mathematical Biology*, Vol. 44, 579–585.

Sonnenschein, Hugo, 1973, "Do Walras' Identity and Continuity Characterize the Class of Community Excess Demand Functions?" *Journal of Economic Theory*, Vol. 6, 345–354.

Svirezev, Yu M., and D. O. Logofet, 1978, *Stability of Biological Relations* (in Russian), Moscow: Nauka.

System Sciences, Inc., 1985, *The Oil Market Simulation Model, Model Documentation Report*, DOE/EI/19656-2, U.S. Department of Energy, Washington, D.C.

Takayama, Akira, 1974, *Mathematical Economics*, New York: Dryden Press.

Tarr, D., 1978, "On Distributed Lags, Morishima Matrices and the Stability of Economic Models," *Review of Economic Studies*, Vol. 45, 1978.

Thomassen, Carsten, 1986, "Sign-nonsingular Matrices and Even Cycles in Directed Graphs," *Linear Algebra and Its Applications*, Vol. 75, 27–41.

———, 1987, "On Digraphs with No Two Disjoint Cycles," *Combinatorics*, Vol. 7, 145–150.

———, 1989a, "Disjoint Cycles in Digraphs," *Combinatorics*, Vol. 11, 393–396.

———, 1989b, "When the Sign Pattern of a Matrix Determines Uniquely the Sign Pattern of the Inverse," *Linear Algebra and Its Applications*, Vol. 199, 27–34.

Todd, John, 1954, "The Condition of Certain Matrices, II," *Archiv Der Mathematik*, Vol. 5, 249.

Topkins, Donald, 1978, "Minimizing a Submodular Function on a Lattice, *Operations Research*, Vol. 26, 305–321.

Tyson, J. J., 1975, "Classification of Instabilities in Chemical Reaction Systems," *Journal of Chemical Physics*, Vol. 62, 1010–1015.

Uzawa, Hirofumi, 1961, "The Stability of Dynamic Processes," *Econometrica*, Vol. 29, 617–631.

Veendorp, E., 1970, "Instability, the Hicks Conditions and the Choice of Numeraire," *International Economic Review*, Vol. 11, 497–505.

Wald, A., 1951, "On Some Systems of Equations of Mathematical Economics," *Econometrica*, Vol. 19, 368–403.

Witte, James, 1966, "Walras' Law and the Patinkin Paradox: A Qualitative Calculus for Macroeconomics," *Journal of Political Economy*, Vol. 76, 72–76.

Yamada, Takeo, 1987, "Generic Matric Sign-Stability," *Canadian Mathematical Bulletin*, Vol. 30, 370–376.

———, 1988, "A Note on Sign-Solvability of Linear Systems of Equations," *Linear and Multilinear Algebra*, Vol. 22, 313–323.

Yamada, Takeo, and Leslie Foulds, 1990, "A Graph-Theoretic Approach to Investigate Structural and Qualitative Properties of Systems: A Survey," *Networks*, Vol. 20, 427–452.

Yamada, Takeo, and T. I. Kitahara, 1985, "Qualitative Properties of Systems of Linear Constraints," *Journal of the Operations Research Society–Japan*, Vol. 28, 331–344.

Yorke, James, and William Anderson, 1973, "Predator-Prey Patterns," *Proceedings of the National Academy of Sciences*, Vol. 70, 2069–2071.

Yun, K., 1977, "Stability of Competitive Industry Equilibrium," *Journal of Economic Theory*, Vol. 16, 177–186.

Zellner, Arnold, and Henri Theil, 1962, "Three-stage Least Squares Simultaneous Estimation of Simultaneous Equations," *Econometrica*, Vol. 30, 54–78.

Name Index

Subject Index